CEN Q & A Handbook Vol IV

CEN Q & A Handbook Vol IV

Mark Boswell

MSN, FNP-C, CEN, CFRN, CTRN, CPEN, TCRN, SCRN, NR-P, W-EMT, EMT-T

Boswell Emergency Medical Education Technology
2018

Copyright © 2018 by Boswell Emergency Medical Education Technology

All rights reserved. This book or any portion thereof may not be reproduced or used in any manner whatsoever without the express written permission of the publisher except for the use of brief quotations in a book review or scholarly journal.

First Printing: 2018

ISBN 978-1-387-43002-4

Boswell Emergency Medical Education Technology
Niceville, FL

www.PassTheCEN.com

Introduction

This is the fourth volume of the CEN Review Handbook series.

This volume continues the same method and outline as set forth in the prior three volumes.

Included in this publishing is a stand-alone 150 question practice CEN-Like exam. Complete answers with rationales are provided in Part 2 of this handbook.

The distribution, skill level, theme, and tone of the questions presented here attempt to mirror the actual CEN exam as much as possible.

The reader or CEN candidate is reminded that it is in repetition of question and answering that the cognitive knowledge base is built and reinforced.

It is my hope that this volume becomes one of the tools to assist in your CEN exam preparation.

As always, feel free to contact me with any feedback, comments, questions or concerns. I strive to be easily reachable by my students and networks.

Good luck!

-Mark Boswell
MSN, FNP-C, CEN, CFRN, CTRN, CPEN, TCRN, SCRN, NR-P, W-EMT, EMT-T

Boswell Emergency Medical Education Technology
www.PassTheCEN.com
facebook.com/bemetweb
BoswellEMT@Gmail.com

Author's Credits

I'd like to thank the following colleagues who participated in the review and editing process. Their efforts are highly appreciated.

Andrew J. Bowman MSN, RN, ACNP-BC, ACNP-C, TNS, CEN, CPEN, CTRN, CFRN, TCRN, CCRN-CMC, CVRN-I-BC, NRP, FAEN
Lebanon, IN

Mindy Lafayette MSN, FNP-BC, CEN, CPEN
Hernando, MS

Katie Keels BSN, RN, CEN
Dallas, TX

George Olschewski BSN, RN, CEN, MICN
Hackensack, NJ

Part 1: Questions

1) Your patient is a 52 year old male with a history of COPD. He is experiencing severe shortness of breath. Blood pressure is 174/106, heart rate is 102, respirations are 32 per minute and shallow/labored. Temp is 99.0. He is slightly drowsy and difficult to remain alert. Which of the following assessments should be done first?

    a) check his SpO2
    b) administer O2 at a low FiO2
    c) prepare for intubation
    d) administer albuterol using air instead of oxygen

2) You are treating a patient who has had a toxic exposure to an organophosphate agent. The airway, breathing, circulation and disability have been assessed and appropriate interventions begun as indicated. Antidote treatment has begun with a dose of atropine and 2-Pam Chloride. Which would be the best physical examination finding to indicate that further administration of the antidote would be expected?

    a) dry mucous membranes
    b) skin hot and dry
    c) tachycardia
    d) miosis

3) During the initial assessment your patient begins actively and violently seizing with generalized tonic/clonic movements. The history is that the patient fell down 1 flight of stairs and hit their head at the bottom. What is the first appropriate action to be done?

a) administer ativan (lorazepam)
b) immobilize the cervical spine
c) begin oxygen at 15 lpm via non-rebreather mask
d) check the pupils for response

4) Which of the following is associated with possible intra-abdominal hemorrhage?

a) Homan's sign
b) Kernig's sign
c) Koplik's spots
d) Cullen's sign

5) Your trauma patient has a confirmed closed, mid-shaft femur fracture. The knee, ankle and hip on the affected side are uninjured. The patient is complaining of severe pain in the mid-femur. There are visible muscle spasms of the thigh area and the dorsalis pedis and posterior tibialis pulses are very faint/weak. A femur traction splint is to be placed. When placed and performed correctly, which of the following is the most important criteria to have accomplished?

a) reduction in patient's pain
b) reduction in the spasms
c) stabilization of the fracture site
d) restoration of the distal pulses to full strength

6) A trauma patient has sustained a significant head injury and has had a ventriculostomy drain placed to monitor and manage intra-cranial pressure (ICP). During reassessment it is noted that the ICP is 30 mmHg and it is found that the stopcock on the drain is open. Which of the following is the best immediate intervention in this situation?

a) elevate the head of the bed to 90 degrees
b) close the stopcock to prevent further drainage of the CSF
c) lower the head of the bed to Trendelenburg position
d) start hyperventilation with 100% oxygen

7) An 18 month old child has sustained a head injury from being dropped from a height. Which of the findings is most consistent with a possible subdural hematoma?

a) decreased ICP measurements
b) conjunctival hemorrhages
c) hypertension
d) bulging fontanelles

8) When considering a possible radiation exposure event, which of the following sets of principles would be appropriate to guide how rescuers can provide care while limiting personal exposure?

a) time, distance, humidity
b) time, distance, shielding
c) distance, shielding, ambient temperature
d) time, shielding and wind direction/speed

9) Which of the following symptoms is most likely associated with a cluster headache?

a) epistaxis
b) aphasia
c) fever
d) lacrimation

10) The patient is a 30 year old male who had both legs pinned between two large pieces of heavy equipment for nearly 3 hours. Which of the following would be an appropriate intervention for this patient?

a) administration of IV potassium chloride to support cardiac function
b) irrigation of open wounds with an antibiotic solution
c) decreasing IV fluid infusion rate to minimize edema and fluid overload
d) infusing isotonic crystalloids to maximize urine output

11) What level of protection should be worn when dealing with a possible biological exposure event containing a highly contagious disease?

a) Level A
b) Level B
c) Level C
d) Level D

12) A patient is being assessed. He was just recently on a scuba diving trip. He complains of dyspnea, chest pain and dizziness. He has a cough which is productive of pink, frothy sputum. Based on this information, which is the best position for this patient to be in?

a) Right lateral decubitus
b) Semi-Fowler's
c) High Fowler's
d) Trendelenburg - Left lateral decubitus

13) Which of the following explains why alkali burns are more serious than acid burns?

a) They cause deep damage to fascia and muscle tissue.
b) They produce liquefaction necrosis.
c) They are typically full thickness.
d) They produce coagulation necrosis.

14) A patient has sustained significant blunt facial trauma. On physical exam the following is noted: a laceration and ecchymosis above the left eye and tenderness, swelling, ecchymosis over the left side of the face. The patient is able to effectively close both eyes, can wrinkle the forehead symmetrically and can elevate the upper lip. There is numbness of the left side of the face. Which cranial nerve has most likely been affected by this injury?

a) facial
b) trigeminal
c) trochlear
d) oculomotor

15) A patient sustained a sudden onset of diffuse urticaria, facial/labial swelling and dyspnea shortly after eating a meal with a friend. Which of the following is the priority intervention?

a) obtain IV access
b) administer high-flow oxygen
c) obtain vital signs
d) administer epinephrine 0.3 mg

16) A patient exhibits recent excessive weight gain, a moon face, truncate obesity, muscle wasting and a buffalo hump. Which condition is most likely the underlying cause for these signs and symptoms?

a) Cushing's syndrome
b) Addison's syndrome
c) Grave's disease
d) SIADH (Syndrome of Inappropriate Anti-Diuretic Hormone)

17) Your intoxicated patient has the following findings: intermittent seizures, decreased deep tendon reflexes, nystagmus and a respiratory rate of 8. Which blood alcohol level is most closely associated with these?

a) 100-249 mg/dl
b) 250-299 mg/dl
c) 300-499 mg/dl
d) >499 mg/dl

18) While assessing a patient, it is observed that they demonstrate a positive Chvostek's sign. What does this indicate?

a) hypercalcemia
b) hypokalemia
c) hyperkalemia
d) hypocalcemia

19) When treating a patient suffering from heatstroke, which of the following would be a complication to watch for?

a) hypomagnesemia
b) hyperkalemia
c) respiratory acidosis
d) metabolic alkalosis

20) Given the following findings: JVD, muffled heart sounds and hypotension, what intervention would be necessary?

a) needle decompression
b) pericardiocentesis
c) peritoneal lavage
d) cricothyrotomy

CEN Q & A HANDBOOK VOL IV

21) You have just assisted in a spontaneous vaginal delivery of a term gestation neonate. There was meconium staining of the amniotic fluid visible at the start of the delivery. Immediate physical exam reveals no spontaneous breathing and limp muscle tone. You are providing warmth and stimulation and the airway appears clear. The heart rate is greater than 100 and there is persistent cyanosis now. Based on the given information, which is the next intervention to be done?

a) assess/perform the APGAR score
b) use the bulb suction to clear the nose then the mouth
c) administer positive pressure ventilation with BVM
d) establish IV/IO access

22) The term "yellow halos" is associated with which possible condition?

a) lithium toxicity
b) cyanide toxicity
c) iron toxicity
d) digoxin toxicity

23) Which of the following is the LEAST likely cause of priapism?

a) sickle cell disease/crisis
b) bacterial infection
c) anticoagulant therapy
d) leukemia

24) Which body part is most likely to develop compartment syndrome when a significant crushing injury has occurred?

a) knee joint
b) upper leg
c) upper arm
d) lower arm

25) Which of the following would be considered to use as initial immobilization for a pelvic fracture?

a) long spine board
b) traction splint
c) pneumatic anti-shock garment
d) bilateral long leg splints

26) Which fracture is most likely to be seen in an adult pedestrian after having been struck by a motor vehicle?

a) forearm fracture
b) femur fracture
c) wrist fracture
d) humerus fracture

27) Which of the following is the most likely potential complication of compartment syndrome?

a) disseminated intravascular coagulation (DIC)
b) myoglobinuria
c) osteomyelitis
d) anemia

28) When is the most likely time following a traumatic event for a fat embolism to occur?

a) within the first 2 hours
b) approximately 4-6 hours
c) approximately 6-10 hours
d) greater than 12 hours

29) Your trauma patient has a confirmed pelvic fracture. Which of the following is the most likely to occur concurrently?

a) rhabdomyolysis
b) urethral injury
c) muscle spasms
d) neurovascular compromise

30) When talking to a grieving family one should:

a) tell the family that the patient didn't suffer
b) use words such as dead, died, or death
c) tell the family, "it was for the best"
d) tell the family, "everything will be all right"

31) Which of the following is true regarding black widow spider bite?

a) The bite may go unnoticed due to minimal pain.
b) The toxin causes a large cholinergic blockade.
c) Male black widows are more likely to bite than females.
d) The most common symptom is a headache.

32) A patient is under your care after having taken a handful of unknown pills and unknown quantity of alcohol. Which of the following is the priority action?

a) determine if there is a history of prior suicidal attempts
b) to determine what support systems the patient has
c) to determine if the patient's condition is life threatening
d) to determine if the patient was trying to harm themselves

33) Which of the following statements is true regarding rabies vaccine administration?

a) A portion of the rabies toxoid is injected around the wound.
b) Rabies immune globulin (RIG) confers long term immunity.
c) All bat injuries/wounds must receive rabies vaccine.
d) Rabies toxoid begins protecting immediately.

34) A 49-year-old woman with a subarachnoid hemorrhage is identified as a potential organ donor. The patient's family consists of her husband, from whom she is separated, elderly parents, her 18-year-old son, and 44-year-old sister. Who is the legal next of kin that can give permission?

a) her son
b) her sister
c) her parents
d) her husband

35) The normal range for intracranial pressure in an adult is:

a) 0 to 15 mm Hg
b) 5 to 30 mm Hg
c) 30 to 45 mm Hg
d) 45 to 60 mm Hg

36) A 3-year-old presents with nausea, vomiting and diarrhea for 36 hours. The assessment shows: pale skin, sunken eyes, tachycardia. The child is unresponsive to pain. The priority intervention for this patient is to administer:

a) intravenous fluids (IVF) immediately
b) oral replacement fluids in small sips
c) Compazine (prochlorperazine) 12.5 mg IM
d) Cipro (ciprofloxacin) 500 mg PO

37) Which of the following signs is most specific for a pulmonary embolus (PE)?

a) heart rate 104 bpm
b) room air SpO2 90%
c) pleuritic chest wall pain
d) hemoptysis

38) An 78-year-old is being discharged after having a laceration of his left leg sutured. Which of the following methods of patient education is most appropriate for this patient?

a) use preprinted instruction sheets
b) prepare handwritten instruction sheets
c) instruct the patient to call his MD if he has any questions
d) conduct a one-on-one discussion with the patient

39) A patient suffering from eclampsia is receiving a magnesium sulfate IV infusion. Which of the following physical findings would most likely suggest magnesium sulfate toxicity?

a) a blood pressure of 128/88 mm Hg
b) an oral temperature of 99.2 F
c) a respiratory rate of 11/min
d) a pulse rate of 90 per minute

40) Your patient has complaints of nasal congestion, sore throat and a nonproductive cough. A chest x-ray has been done and read as negative. Based on the given information the most likely diagnosis is:

a) pneumonia
b) acute bronchitis
c) asthma
d) COPD

41) Assuming all the assessments and interventions are warranted, which of the following would be done in the correct sequence?

a) assist ventilation with BVM, assess skin temperature, suction airway, start 2 large bore IVs
b) assess skin temperature, start 2 large bore IVS, attach cardiac monitor, suction airway
c) attach cardiac monitor, assist ventilation with BVM, assess skin temperature, start 2 large bore IVs
d) suction airway, assist ventilation with BVM, start 2 large bore IVS, attach cardiac monitor

42) You are caring for a 20 year old male who sustained a fall of approximately 20 feet from a rooftop. He is not responding to your voice as you approach to begin patient care. His eyes are closed and no spontaneous movement is noted. Which of the following is the first priority?

a) perform a jaw-thrust and assess the airway
b) perform a head-tilt, chin lift and assess the airway
c) check for a carotid pulse
d) start positive pressure ventilation with BVM if RR <10

43) A victim of domestic violence is being cared for. At the time of the acute evaluation and treatment, which is the most important thing to assess for?

a) history of previous victimization
b) concurrent use of drugs or alcohol
c) the readiness to leave the perpetrator
d) the reasons why they stay in an abusive relationship

44) SIDS (Sudden Infant Death Syndrome) is one of the common causes of death in the early childhood period. Which age group has the highest risk of SIDS?

a) 12-24 months
b) 6-12 months
c) 1 week - 1 year
d) 6-8 weeks

45) Your patient has been diagnosed with community acquired pneumonia (CAP). Continuous pulse oximetry is showing a decline into the mid-high 80s. Given this information which of the following is the most accurate statement?

a) There is an oxygenation problem.
b) There is a ventilation problem.
c) The patient is hypoxic.
d) The patient is hypercarbic.

46) Which of the following is a possible result of multiple packed red blood cell transfusions?

a) hypercalcemia
b) hypocalcemia and hypothermia
c) hypothermia and hyperchloremic acidosis
d) hyponatremia

47) The patient is a 44 year old male complaining of chest pain. Just as he is placed on the cardiac monitor, he gasps and becomes suddenly unresponsive. The monitor shows a wide complex tachycardia. Which of the following is the most likely interpretation?

a) supraventricular tachycardia
b) idioventricular
c) atrial fibrillation
d) ventricular tachycardia

48) While treating your patient, you administer 100 mcg of Fentanyl via IV push. The patient suddenly becomes unable to breathe. Immediately the patient is ventilated with 100% O2 by bag-valve-mask. It is suddenly noted that there is no chest rise and fall the chest has become rigid and will not expand with ventilations. Which of the following will initially be ordered to begin managing this life threatening complication?

a) narcan
b) romazicon
c) succinylcholine
d) lidocaine

49) Which of the following is a contraindication to the administration of succinylcholine?

a) hypernatremia
b) glaucoma
c) reactive airway disease
d) a recent concussion

50) For which of the following would cardioversion be contraindicated?

a) digoxin toxicity
b) hypermagnesemia
c) cyanide toxicity
d) methemoglobinemia

51) For which of the following is it most important to receive appropriate antibiotics?
a) a 2 cm scalp laceration from a rusty nail
b) acute sinusitis
c) strep pharyngitis
d) a cat bite wound

52) A patient is only responsive to pain stimulus. The cardiac monitor shows a narrow complex waveform at a rate of 44. The history confirms the patient is a heart transplant recipient. Which drug should be considered?

a) adenosine
b) epinephrine
c) isoproterenol
d) atropine

53) Which of the following is a late sign of hypovolemic shock in the pediatric patient?

a) tachycardia
b) markedly decreased bilateral breath sounds
c) diaphoresis
d) hypotension

54) A dopamine IV infusion is currently running. The insertion site appears to show that the dopamine IV has extravasated. Which of the following would be used to counteract potential soft tissue necrosis?

a) calcium gluconate
b) phentolamine (Regitine)
c) sodium thiosulfate
d) hydralazine (Apresoline)

55) In which of the following scenarios is dopamine contraindicated?

a) hypotension due to a tachyarrhythmia
b) a concurrently running dobutamine IV
c) hypovolemic hypotension following fluid resuscitation
d) normo-volemic hypotension

56) A lidocaine IV infusion is ordered at 3mg/min. The lidocaine is mixed as 1 gram in 250 ml of D5%W. What would be the correct rate to run this at with a 60 microdrip IV tubing set?

a) 75 ml/hr
b) 45 ml/hr
c) 30 ml/hr
d) 15 ml/hr

57) A dobutamine infusion is ordered at 6 mcg/kg/min for a 75 kg patient. A 60 gtt drip set will be used. The drug mixture is 500mg in 250 ml D5%W. At what rate should it run to deliver the ordered dose?

a) 15 ml/hr
b) 14 ml/hr
d) 12 ml/hr
d) 11 ml/hr

58) Which of the following is true about magnesium sulfate?

a) used in the treatment of atrioventricular block
b) used to promote diuresis for patients with renal insufficiency
c) increases the tone of smooth and skeletal muscle tissue
d) acts to depress the central nervous system (CNS)

59) Which of the following is true regarding the preparation and administration of an IV infusion of nitroglycerin?

a) it should be diluted in D5W and in a glass bottle
b) PVC tubing should be used
c) filtered IV tubing should be used
d) the IV container must be wrapped in an opaque material to prevent light exposure

60) A patient has overdosed on a large amount of an antihistamine and is in an anti-cholinergic crisis. You would expect to receive a drug order for which of the following?

a) Dobutrex (dobutamine)
b) physostigmine
c) Inotropin (dopamine)
d) Narcan (naloxone)

61) Given only the conditions listed below, which would be the highest triage priority?

a) congestive heart failure
b) chest pain
c) seizures
d) severe abdominal pain

62) A 2 year old child is suspected of having Reye's Syndrome. Which set of findings is most specific for this?

a) a red, raised, "bull's eye" shaped rash, multiple joint pains and malaise
b) peripheral edema, a fever for 5 days and a "strawberry tongue"
c) fever, depressed level of consciousness and altered liver function tests
d) multiple joins inflamed, a red macular rash with a clear center and a low grade fever

63) Which finding is most likely associated with a patient having acute respiratory distress syndrome (ARDS)?

a) increased HCO3
b) elevated PaCO2
c) decreased PaO2
d) elevated COHb

64) Your patient is a 43 year old male. He was at work at a local chemical plant. There was a spill of some toxic vapors of which he inhaled some of them. He is having shortness of breath, pleuritic chest pain and an intractable cough. Pulse is thready at 122. Blood pressure is 108/76. Respirations are 30 and unlabored. Temp is 98.0. Lung sounds are clear, skin is pink, capillary refill is brisk. His chest x-ray shows bilateral infiltrates and a normal heart size. A BNP level is drawn and it is normal. Based on this information alone, which of the following would you expect to be ordered first?

a) Solumedrol IV
b) racemic epi via nebulizer
c) magnesium sulfate IV
d) thiamine IV

65) Which test will most accurately identify a pulmonary embolus (PE)?

a) bronchoscopy
b) pulmonary angiography
c) chest x-ray
d) ventilation/perfusion (V/Q) scan

66) When considering pulmonary embolus as the differential, which of the following patients would benefit from the use of the D-Dimer test?

a) 26 year old female who smokes cigarettes
b) 35 year old male, heart rate 104
c) 52 year old female, left leg is swollen and tender
d) 46 year old female, taking estrogen

67) The caregiver of a 4 year old reports the child has had a sudden onset of high fever, cough, sore throat, irritability and "funny sounds" when breathing. Rapid assessment reveals a sick appearing child, who is sitting up in a tripod position, with back and neck arched up while breathing. Temperature is 103.2, pulse is 152, respirations are 30 and stridor is heard from the airway. The child appears tired and fatigued. Drooling is noted as well. The provider elects to get a lateral neck x-ray. Which of the following x-ray findings would be expected to be seen given this presentation?

a) thumbprint sign
b) steeple sign
c) supraglottic narrowing
d) subglottic narrowing

68) A 32 year old male sustained a fall from a 20 foot ladder. He is complaining of pain in the mid/lower thoracic spine. Your patient assessment reveals the following: Core temp 99.0, Respirations 24, Pulse 62, Blood pressure: 76/46. He is A/O x 3. Distal motor and sensory is intact in all extremities. He is pale, clammy and diaphoretic above the mid thorax and his lower extremities are pink, warm and dry. Based on these findings, which of the following would be expected to be ordered FIRST?

a) administer atropine IV
b) start a dopamine infusion
c) administer a 1 liter colloid bolus IV
d) prepare for cardiac pacing

69) Which of the following would be the most serious injury possibly associated with a fracture of the first or second rib?

a) clavicle fracture
b) tracheal tear
c) aortic rupture
d) cervical spine injury

70) Your patient has sustained a traumatic blow to the left thoracic wall. There are rib fractures suspected. This patient is also at risk for a pulmonary contusion. Which of the following findings is strongly tied to poor outcomes from the pulmonary contusion?

a) bilateral crackles in the lung bases on auscultation
b) hemoptysis
c) temperature of 100.4 F
d) a WBC count of 30,000

71) Which of the following is true regarding tetanus vaccination?

a) Tetanus toxoid is an active vaccination.
b) Pregnancy is a contraindication to tetanus toxoid administration.
c) HIV is a contraindication to tetanus toxoid administration.
d) Tetanus immune globulin is good for 5-10 years.

72) A 2 year old child presents with a sudden onset of severe abdominal pain. It started suddenly during normal activity. The child has been in his usual state of health recently. The pediatric assessment triangle (PAT) reveals: child is in marked distress and crying, respirations are somewhat rapid and slightly shallow; there is no suggestion of an airway compromise, the skin appears somewhat pale and moist. The clinician is strongly considering the condition to be a bowel intussusception. Which of the following symptoms has most likely lead the clinician to consider this?

a) the presence of currant jelly stool
b) diarrhea x 24 hours
c) a fever > 102
d) projectile vomiting

73) Which of the following accounts for the majority of deaths associated with chest trauma?

a) motor vehicle collisions
b) firearms
c) assaults
d) falls

74) A patient with COPD receiving high levels of O2 is likely to experience which of the following?

a) a marked drop in the CO2
b) a ventilation/perfusion mismatch
c) decreased ventilatory drive
d) decreased vital capacity

75) While treating a patient who has sustained a prolonged submersion injury and is hypothermic, you recall which of the following should be anticipated during their initial care?

a) Pt will be in an alkalotic state.
b) Routine ACLS/resuscitation guidelines will be followed.
c) Shivering will still be present at 85 degrees F.
d) Pupils will be constricted.

76) During which stage are the classic findings of Pertussis most likely to be observed?

a) catarrhal
b) paroxysmal
c) coryzal
d) convalescent

77) A patient has sustained an open pneumothorax. Immediately a non-porous dressing was placed over the open wound. On reassessment which of the following would be associated with a worsening of the patient's condition?

a) oxygen saturation of 93%
b) decreased breath sounds on the affected side
c) blood pressure of 110/70
d) tracheal shift and jugular venous distention

78) Which of the following is the most common cause of pediatric cardiac arrest?

   a) a cardiac arrhythmia
   b) hypoxia
   c) an electrolyte disturbance
   d) a congenital cardiac abnormality

79) Which of the following are the most common causes of acute pancreatitis?

   a) trauma and alcohol abuse
   b) gallbladder disease and alcohol abuse
   c) substance abuse and hypercalcemia
   d) post-operative complications and trauma

80) Your patient has ingested an alkaline chemical and sustained tissue damage to the esophagus and upper gastrointestinal tract. Which of the following interventions would be a part of the management of this patient?

   a) administer steroids
   b) induce vomiting
   c) insert a nasogastric tube (NG)
   d) manage the airway

81) A patient has just been diagnosed with tuberculosis and an order to give isoniazid (laniazid) is given. This drug would be contraindicated if which of the following were present?

   a) patient is already taking furosemide (lasix)
   b) the patient has glaucoma
   c) the patient has known coronary artery disease
   d) the patient is currently taking dilantin (phenytoin)

82) A pheochromocytoma is most often first diagnosed in which age group?

a) under age 10
b) ages 25-30
c) ages 31-60
d) over age 60

83) Which of the following is most likely (most specific) to be seen in association with a mesenteric infarction?

a) vomiting
b) bloody diarrhea
c) free air under the diaphragm
d) constipation

84) In the patient with hypernatremia, which of the following will be a primary treatment?

a) administering activated charcoal
b) administering Kayexalate
c) administering diuretics
d) administering fluids

85) The patient is a 3 year old African American male child. He is inconsolable, crying, and irritable. His rectal temperature is 101.9 F. On physical exam it is noted he has swollen hands predominantly over the joints. Based on this information, which do you suspect?

a) sickle cell crisis
b) meningitis
c) discoid lupus flare
d) hepatitis

86) A 24 year old paraplegic patient is being evaluated. He complains of a severe headache. Vital signs are: BP = 236/112, HR = 62, RR = 26 and temp = 99 deg F. The patient is anxious and distressed. An indwelling foley catheter is inserted and 100 ml of urine is drained. It is additionally noted that this patient has several decubitus ulcers. Which of the following medications would most likely be administered first?

a) heparin
b) Demerol (meperidine)
c) Calan (verapamil)
d) Procardia (nifedipine)

87) A 33 year old male is complaining of scrotal pain for the past day. Which of the following findings would help distinguish epididymitis from testicular torsion?

a) an elevated creatinine
b) a urinalysis showing bacteriuria
c) a CBC showing leukopenia
d) hypoperfusion on testicular doppler ultrasound

88) A patient has sustained a possible stroke and is demonstrating receptive aphasia. Which area of the brain has most likely been damaged?

a) temporal lobe
b) occipital lobe
c) parietal lobe
d) frontal lobe

89) Which of the following sets of symptoms is most consistent with pregnancy-induced hypertension (PIH)?

a) proteinuria, double vision, uterine contractions
b) proteinuria, headaches, double vision
c) headaches, vaginal bleeding, proteinuria
d) double vision, vaginal bleeding, headaches

90) A patient is 33 weeks pregnant. Her blood pressure is 182/112 mmHg. She is complaining of a headache and blurred vision. Which of the following should you be most concerned about happening?

a) seizure activity
b) cardiac arrhythmias
c) vaginal bleeding
d) precipitous delivery

91) For the patient diagnosed with iritis, which of the following would be the most specific and appropriate treatment?

a) Refer the patient immediately to an ophthalmologist.
b) Administer topical decongestants, 1 drop 3-5 times daily.
c) Administer prednisolone and tropicamide (Mydriacyl).
d) Administer erythromycin.

92) Which of the following interventions is a priority for the patient with a retinal detachment?

a) Prepare the patient for emergent surgery.
b) The patient should be referred to an opthamologist.
c) Both eyes should be patched.
d) The patient should be placed on immediate bed rest.

93) A patient who recently sustained a mid-femur fracture is having symptoms suggestive of fat embolism syndrome. Which of the following is a primary intervention?

a) administration of high-flow oxygen
b) administration of an isotonic IV fluid bolus, to maintain preload and cardiac output
c) administration of corticosteroids
d) placing the patient in trendelenburg position

94) Which of the following statements is true regarding the care of an amputated part?

a) Anticipate the need to x-ray the amputated part before reimplantation.
b) The amputated part will be discarded if it has been longer than 60 minutes.
c) The amputated part should be immersed in a solution of ice and water.
d) The amputated part should be wrapped in a towel and moistened with sterile saline.

95) Which of the following is most likely associated with a nerve entrapment syndrome?

a) lumbar vertebral column
b) thoracic vertebral column
c) the leg
d) the arm

96) Which of the following groups of injuries are most likely associated with injuries to unrestrained front seat passengers involved in a motor vehicle collision?

a) patellar fracture, foot fracture, hip fracture
b) ankle fracture, hip dislocation, foot fracture
c) hip dislocation, patellar dislocation, ankle fracture
d) ankle dislocation, patellar dislocation, hip dislocation

97) In the Emergency Department setting, which of the following is the most appropriate intervention to control hemorrhage associated with an extremity injury?

a) apply a pressure dressing to the bleeding site
b) applying cool compresses to limit bleeding by vasoconstriction
c) placing a tourniquet 3-4 inches above the bleeding site
d) holding the extremity in a dependent position and applying pressure

98) A patient experiencing an acute psychosis has just received an appropriate antipsychotic and is beginning to experience a dystonic reaction. Which medication would be expected to be ordered to offset this reaction?

a) midazolam (Versed)
b) prochlorperazine (Compazine)
c) diphenhydramine (Benadryl)
d) haloperidol (Haldol)

99) Which of the following are considered risk factors for bipolar disorder?

a) multiple divorces, legal/financial troubles, age of onset in 50's
b) legal/financial troubles, age of onset in the 40's, more likely female
c) age of onset in the 40's, legal/financial troubles, multiple divorces
d) hypermanic episodes, psychosis, age of onset 40's

100) In a toddler, which of the following injury patterns is most likely to be associated with child abuse?

a) a bruise on the right shin
b) a 2.4 cm laceration to the forehead
c) a hematoma over the occiput
d) several 1-1.5 cm diameter circular burns on the posterior thorax

101) For the acutely psychotic patient requiring medication, which of the following would most likely be considered?

a) amitriptyline (Elavil)
b) haloperidol (Haldol)
c) chlorpromazine (Thorazine)
d) lithium (Lithonate)

102) Post-traumatic stress reactions are most likely to be exhibited by which of the following?

a) flashbacks, recurring dreams and numbness
b) denial of the events
c) anger, guilt and humiliation
d) fatigue and self blame

103) A 42 year old female is complaining of her heart racing hard and also being scared about having an acute MI. She further explains that she is always worried about her personal safety and concerns about a natural disaster is looming. She also reported difficulty sleeping and feeling "edgy" for about 4 months. Which of the follow statements is most reflective of this scenario?

a) This patient's behavior suggests substance abuse.
b) This patient is exhibiting possible anxiety or panic symptoms.
c) This patient is malingering and attention seeking.
d) This patient may be in the early stages of Alzheimer's.

104) Your patient is experiencing chest pain. They have a sustained release nitroglycerin topical patch in place on the chest wall. During interview and monitoring they go into v-fib. As you prepare to defibrillate them, you recall that the nitroglycerin patch needs to be removed because:

a) it may cause a reflex tachycardia following defibrillation
b) electrical arcing may occur
c) the drug may become neutralized or deactivated
d) defibrillation may cause too much of the drug to be released

105) In which of the following scenarios would respiratory acidosis be a possible finding?

a) high altitudes
b) aldosteronism
c) severe scoliosis
d) salicylate toxicity

106) Which of the following is the best statement regarding critical incident stress?

a) It is seen only in the direct caregiver(s).
b) It is defined by the individual.
c) It is seen as occupation-related stress.
d) It is seen frequently throughout the careers of emergency providers.

107) An EKG is performed on a patient and it shows ST-segment elevations in leads II, III and aVf. This indicates likely occlusion of which artery?

a) right coronary
b) left coronary
c) left anterior descending
d) circumflex

108) A patient is to receive an IV infusion of nitroglycerin. It is ordered to be run at 25 mcg/minute. The drug is prepared as 100mg of nitroglycerin in 250 ml of D5W. Using a 60 gtt/ml infusion set, what would the infusion rate be?

a) 4 ml/hr
b) 6 ml/hr
c) 8 ml/hr
d) 10 ml/hr

109) The patient is to receive an IV infusion of dopamine at 10 mcg/kg/min. The patient weighs 80 kg. The dopamine IV is mixed as 400 mg in 250 ml of D5W. What will the infusion rate be using a 60 gtt/ml infusion set?

a) 25 ml/hr
b) 30 ml/hr
c) 35 ml/hr
d) 40 ml/hr

110) Which of the following drugs has the desired effect of increasing cardiac response to catecholamines?

a) atropine
b) sodium bicarbonate
c) lidocaine
d) dopamine

111) Which of the following has the highest priority to undergo a noncontrast head CT scan following a closed head injury?

a) pt is intoxicated, without other findings
b) pt reports persistent dizziness
c) pt reports nausea
d) pt was confused for 5 minutes following the injury

112) Which of the following is the most accurate description of a flail chest?

a) a fracture of two or more ribs in two places
b) a fracture of two adjacent ribs, bilaterally
c) a compressed rib cage and an open chest wound
d) a segment of the chest wall that moves paradoxically with respirations

113) One of the possible complications of a pulmonary embolus is right sided heart failure. Which of the following would be most consistent with this?

a) the presence of a pericardial friction rub
b) an expiratory wheeze
c) a peaked P wave on the EKG
d) a physiologic S2 split heart sound

114) Which of the following is the highest priority for a patient with a pulmonary embolus?

a) administration of an analgesic for the chest pain
b) correcting the hypoxia with supplemental oxygen
c) administering heparin at a IV dose of 1,000-3,000 units/hour
d) administering a thrombolytic

115) A trauma patient has had a central line placed and their CVP is reading 3 cm H2O. Which of the following statements reflects the significance of this value?

a) The patient has a fluid volume deficiencey.
b) The patient has an alteration in their gas exchange.
c) The patient has an excess of fluid volume.
d) The patient is hypoxic.

116) What result would be seen with alpha-adrenergic receptor stimulation?

a) vasoconstriction
b) initially a vasodilation, then a vasoconstriction
c) tachycardia
d) vasodilation

117) In Class I Shock, which of the following would be expected?

a) increased systolic and diastolic pressures
b) normal or decreasing systolic pressure and rising diastolic pressure
c) rising systolic pressure and falling diastolic pressure
d) falling systolic and diastolic pressures

118) In the pediatric patient, which of the following findings is most consistent with late stages of shock?

a) irritability
b) urine output of 1-2 ml/kg/hr
c) tachycardia
d) bradycardia

119) For which of the following types of shock will dopamine be the least useful and possibly harmful?

a) neurogenic
b) cardiogenic
c) septic
d) hypovolemic

120) A patient is brought in and is suspected of methanol toxicity. Which of the following is a realistic concern in this setting?

a) reduced cardiac output
b) altered mental status
c) ineffective ventilations
d) altered perceptions/hallucinations

121) Which of the following is most likely associated with an amphetamine overdose?

a) miosis
b) hot, dry skin
c) tachycardia
d) hypotension

122) Following a potentially venomous snake bite the decision to administer CroFab is given. However, it will take 4 hours for the meditation to be prepared. In the meantime, what should be done?

a) Immobilize the injured extremity.
b) Apply a tight constricting band proximal to the envenomation site.
c) Apply ice to the envenomation site.
d) Elevate the affected extremity above the level of the heart.

123) What is the primary factor to consider when making the decision to transfer a patient to another facility?

a) patient's insurance status
b) consent of the patient or family if patient unable
c) the mode of transportation
d) the risks and benefits to the patient

124) A patient is being taught about home care during their discharge. They are being taught about which adverse conditions should be reported to their doctor. What kind of learning/teaching style is this?

a) social
b) cognitive
c) affective
d) psychomotor

125) What's the most common unintentional tort involving health care personnel?

a) negligence
b) malpractice
c) battery
d) assault

126) Several sources of law affect emergency health care workers. What source of law would Medicare laws fall under?

a) statutory law
b) ordinances
c) common law
d) constitutional law

127) In the patient who has had a heart transplant in the past, what is the usual presenting symptom for an acute myocardial infarction?

a) jaw pain
b) substernal chest pain
c) heart failure symptoms
d) tachycardia

128) A patient is taking prednisone 10 mg daily for lupus. They have been taking it for many years. If they suddenly discontinue it, which of the following is most likely to be caused?

a) tardive dyskinesia
b) psychiatric disturbances
c) adrenal crisis
d) glaucoma

129) For the patient recently diagnosed with a brain tumor, what is the purpose of corticosteroids?

a) to provide analgesia
b) to reduce cerebral edema
c) to treat agitation
d) to control or prevent absence seizures

130) A patient is being treated for an intentional aspirin overdose. Which of the following is most likely to be added to their IV infusion that is running?

a) magnesium
b) sodium bicarbonate
c) folic acid
d) calcium gluconate

131) Which of the following is the most often reported emotion related to critical incident stress?

a) depression
b) hebephrenia
c) euphoria
d) panic

132) A trauma patient has sustained a hemothorax from blunt chest injuries. A chest tube has been placed on the left side succesfully. On initial insertion 600 ml of blood was drained. Following the thoracostomy the vital signs showed a heart rate of 128, a respiratory rate of 24, and a blood pressure of 138/72. While awaiting admission to the trauma ICU, which of the following reassessments is the most important over the next hour?

a) urine output
b) chest tube drainage
c) central venous pressure
d) vital signs

133) Which of the following is least likely associated with a flail chest?

a) pulmonary contusion
b) respiratory distress
c) sucking chest wound
d) paradoxical chest wall movement

134) A pediatric patient is under your care. They look "sick," with labored breathing, drooling, and a decreased level of consciousness. On exam it is noted the child has a fever of 102.4 F, heart rate of 128, respirations of 30. There stridor of the upper airway and the child is restless and attempting to sit up straight. The caregiver relates that the child was in his normal state of health recently and this all just began in the last 12 hours. Based on your suspicion of the protential problem, which of the following is the priority for this?

a) applying the ECG to watch for arrhythmias
b) assisting with intubation
c) administering IV antibiotics
d) applying oxygen via face mask

135) A 50 year old male presents for care. He is currently undergoing chemotherapy treatments for prostate cancer. His symptoms include: fever, nausea and vomiting. Which of the following is the appropraite 5-level ESI category for him?

a) Level 1
b) Level 2
c) Level 3
d) Level 4

136) An endotracheal tube has just been placed as part of the treatment for a tracheobronchial injury. Which of the following suggests the intervention was effective?

a) improved ABG panel
b) jugular venous distention
c) tracheal shift
d) respirations are 36 per minute

137) A patient is being treated for septic shock. Which of the following is most suggestive of a poor outcome?

a) CVP 4 cm H2O
b) ABG pH 7.35
c) serum lactate 6 mmol/L
d) blood pressure 92/56 mm Hg

138) While resuscitating a trauma patient suffering from hemorrhagic (hypovolemic shock), in addition to PRBCs, what other product should be considered for administration?

a) washed RBCs
b) FFP or platelets
c) Dextran
d) albumin

139) After performing initial fluid resuscitation to a patient in septic shock with 3,000 ml of NS, the urine output is 35 ml/hr. Which of the following is the next appropriate intervention?

a) adminster another 1,000 ml NS by rapid bolus
b) continue NS infusion at 125 ml/hr
c) change the IV to D5W and give 1,000 ml bolus
d) give another 2,000 ml bolus

140) Law enforcement has requested an evaluation for a patient who was apprehended in posession of cocaine. They report that during the arrest the patient ingested the substance. What finding should be observed for to indicate the possibility of cocaine overdose?

a) tachycardia
b) miosis
c) hypothermia
d) hypotension

141) A young man is being treated following an accidental ingestion of gasoline while siphoning it from a vehicle. Which of the following is going to be the priority assessment?

a) cardiac
b) neurological
c) respiratory
d) gastrointestinal

142) A male presents with complaints of headache, low grade fever, abdominal cramping and myalgias. Symptoms began gradually approximately 7-8 days following a week long camping trip in North Carolina. On exam, a deep-red rash is noted on the soles of the feet, the wrists and the ankles. It appears petechial and purpuric. Which of the following is the most likely cause for this patient's illness?

a) measles
b) poison ivy
c) Rocky Mountain Spotted Fever
b) Lyme disease

143) When considering a choice of transportation for transferring a patient, what advantage does ground transport have over helicopter transport?

a) fewer traffic and road factors
b) faster speed
c) more patient care space
d) better radio communications

144) When considering methods to provide patient education, one must realize certain methods take longer to accomplish. Which of the following types of learning takes the longest?

a) affective
b) social
c) psychomotor
d) cognitive

145) A working group of RNs, MDs, ER techs and registration personnel are attempting to work on a solution to a long length of stay in the ED. Their initial process included gathering data on the differences in length of stay based on patient acuity levels. This data gathering is an early step in which of the following processes?

a) quality improvement process
b) collaborative research
c) indicator relevance testing
d) descriptive research

146) If an ED has implemented a shared governance model, which of the following is most likely to be utilized?

a) computer generated scheduling
b) a lead staff scheduler
c) sheduling done by management
d) self-scheduling

147) Which of the following conditions is most commonly associated with bradycardia and AV node conduction problems?

a) heart failure
b) inferior MI
c) anterior MI
d) cardiogenic shock

148) A patient has suffered a 20' fall from a scaffold. He complains of pain in his chest through to his upper back. The chest x-ray returns with the interpretation of a widened mediastinum. What is the most likely cause for this?

a) ruptured aorta
b) ruptured trachea
c) pneumothorax
d) ruptured hemidiaphragm

149) A patient complains of blurred vision and pain in the right eye. Examination shows a cornea that is hazy on that side as well as a pupil that is irregular and sluggish to react. Which of the following is the most likely cause for this?

    a) central retinal artery occlusion
    b) iritis
    c) glaucoma
    d) conjunctivitis

150) A patient has sustained blunt injury to the abdomen. They are exhibiting hypotension, tachycardia, absent bowel sounds, an abdomen that is firm and tender to palpation and visible bruising and abrasions over the left and right upper quadrants. Which of the following most likely explains these findings?

    a) diaphragmatic rupture
    b) small bowel contusion
    c) liver or spleen laceration
    d) aortic rupture

Boswell

Part 2: Answers & Rationales

1) ANSWER: a) check his SpO2
This patient is experiencing a degree of respiratory distresss. The question is asking for which "assessment" should be done first. To answer this question you must consider which possible answers are "assessments". Checking the SpO2 is the only answer that is an assessment. All the other answers are interventions. While one may consider that this patient needs some immediate intervention, recall that the question is looking for an assessment, not an intervention.

2) ANSWER: d) miosis
Treatment with atropine (an anticholinergic) and 2-Pam Chloride are the antidotes for organophosphate poisoning. Organophosphate poisoning would exhibit constricted pupils (miosis), so the continued presence of miosis indicates the additional need for antidote administration. Organophosphate poisoning would cause copious amounts of secretions from the skin (sweat) and the mucous membranes (salivation) as well as lacrimation. Therefore the presence of dry mucous membranes and hot/dry skin indicate that the antidote is effective. Tachycardia is an expected side effect of atropine administration.

3) ANSWER: c) begin oxygen administration at 15 lpm via non-rebreather mask

Following the priorities of assessment beginning with the ABC's, addressing the airway with simultaneous c-spine immobilization is first. In this case, with the patient actively seizing, immobilizing the c-spine would prove difficult and possibly detrimental (maybe causing injury). The next consideration in sequence would be addressing the breathing. Application of a non-rebreather mask is something that could be done even with a seizing patient and the benefit would be justified. Compared with cervical spine immobilization which might cause injury, application of a non-rebreather should cause no injury. Checking the pupils may be considered, however this would come under the "D" assessment (disability) which is after ABC. Ativan (lorazepam) would be one of the preferred agents for use during a seizure, however, addressing the ABCs in order is the correct sequence.

4) ANSWER: d) Cullen's sign

Cullen's sign is bruising or ecchymosis visible at the periumbilical area. It may be suggestive of intra-abdominal hemorrhage or pancreatitis. When accompanied by Grey-Turner's sign (flank ecchymosis) it is much more specific for intra-abdominal hemorrhage. Homan's sign is associated with assessment for a possible DVT (deep vein thrombosis). It is elicited by causing sudden dorsiflexion of the foot on the suspected side and is considered positive if it causes pain in the calf. It is not commonly used as it has poor predictive value. Kernig's sign is useful to check for meningitis which causes meningeal irritation. It is elicited by flexing the hip to near 90 degrees and then extending the same knee. This induces strain and increased pressure on the meninges. A positive response will have the patient reflexively flex the neck in an attempt to decrease the pressure. Koplik's spots are near pathognomonic for measles. They are seen inside the buccal (cheek) mucosa.

5) ANSWER: c) stabilization of the fracture site
The indication for application of a femur traction splint is a mid-femur fracture. The purpose of applying it is to stabilize the fracture site. This attempts to prevent further injury or damage as well as to reduce or tamponade the hemorrhage which usually accompanies this injury. Reduction in pain, reduction in spasms and increasing distal blood flow all may occur as a result of proper application, however, they are not the first/foremost purpose for application. In other words, even if the patient's pain does not improve, nor the spasms decrease nor the distal pulses increase; as long as the fracture site has been stabilized - the purpose has been achieved. The traction placed on this injury helps to also tamponade the thigh compartment thereby decreasing or slowing hemorrhage which may be occurring in that area. Some references cite a possible 1.5-2.0 liters of blood lost into the thigh area.

6) ANSWER: d) start hyperventilation with 100% oxygen
The patient is at risk for herniation and immediate intervention is required to help minimize this. Hyperventilation will drive the CO_2 down and cause vasoconstriction which will reduce the cerebral blood volume, thus helping to decrease ICP. Lowering the head of the bed to Trendelenburg position would cause the ICP to increase further. Elevating the head of the bed would help reduce the ICP, however, it will also decrease perfusion - this is not the ideal, best answer. Closing the stopcock would effectively cause the ICP to raise as there would be no outlet for the pressure.

7) ANSWER: d) bulging fontanelles
In the child younger than 2 years old, bulging fontanels would be highly suspicious for a possible subdural hematoma. The incompletely fused skull bones would allow for expansion of the intracranial compartment and appear bulging at the fontanel spaces. Increasing ICP (related to a subdural hematoma) would more likely cause retinal hemorrhages, not subconjunctival hemorrhages. Hypertension would be a very non-specific finding and is associated with many potential etiologies. The ICP would be expected to increase, not decrease, in a subdural hematoma.

8) ANSWER: b) time, distance, shielding
When considering radiation exposures, the gamma and neutron rays are of the most serious concern, as their wavelengths permit them to travel through human tissues. The most important principles to limit radiation exposure are time, distance and shielding. The amount of time in close proximity to the source will increase the amount of exposure. The closer to the source will increase the total dose exposed to. The more shielding between the source and the patient will decrease the amount of exposure. Humidity, ambient temperature and wind direction/speed are not as influential as time, distance, shielding.

9) ANSWER: d) lacrimation
Typical symptoms associated with cluster headaches are: excruciating, unilateral pain, usually behind the eye or in the temporal area. They may have tearing, nasal congestion and facial flushing and/or rhinorrhea. Epistaxis is not associated with cluster headaches. Aphasia would be more specific to a stroke or complex migraine. Fever is a non-specific finding.

10) **ANSWER: d) infusing isotonic crystalloids to maximize urine output**
This patient has sustained a concerning crush injury. The patient will be at risk for rhabdomyolysis due to the accumulation of myoglobin from the damaged tissues. Potassium levels may also be elevated due to the crushed/damaged tissues. The best answer choice is to increase the IV fluid rate based upon urine output. This will ensure adequate filtration through the kidneys and prevent rhabdomyolysis. Administering IV KCL would be contra-indicated due to the potential of an already elevated potassium level from the crushed tissues. Irrigating open wounds with an antibiotic solution would not be incorrect, however, this is not the priority at this point. Decreasing IV fluids would be counter-productive and might increase the likelihood of ensuing rhabdomyolysis.

11) **ANSWER: c) Level C**
Level C protection includes a tyvek suit, gloves and a purified positive air respirator (PAPR). This is recommended for many reasons. The PAPR is easier to use and can be used for longer periods of time than the higher level devices. A level B respirator has a limited air supply thus limiting the useful time for that device. Level A and B protections require the user to be completely enclosed in an encapsulated suit - which is not necessary for a biological event. Level D protection is routine work clothing and provides no respiratory protection.

12) ANSWER: d) Trendelenburg - Left lateral decubitus
This patient is most likely experiencing an air embolism commonly caused by failure of the diver to exhale during ascent from depth. Place the patient in Trendelenburg-Left lateral decubitus position (Durant's maneuver). This position maximizes the chances of keeping the air embolus in the right atrium and not entering the pulmonary circulation. In the patient who may have a patent foramen ovale, this prevents the air embolus from crossing to the left side and becoming a potential cerebral embolus.

13) ANSWER: b) They produce liquefaction necrosis.
Liquefaction necrosis is the predominant histological tissue damage pattern caused by alkaline substances. They cause rapid damage caused by neutralization of the substance leading to the release of thermal energy, edema, inflammation, thrombosis, cell death, and ultimately necrosis. Coagulation necrosis is the predominant histological tissue damage pattern caused by acids. They cause slower damage from activation of the coagulation cascade, thrombosis, ischemia, and ultimately necrosis; acids do have high tissue penetration (although slower). While full thickness damage, in general, is deleterious, it alone is not the major factor in the alkaline burns being more severe. Extensive muscle and fascial tissue damage is most often associated with electrical burns.

14) ANSWER: b) trigeminal
The trigeminal nerve covers facial sensory information and motor control for jaw movement. The oculomotor and trochlear nerves are motor only. They control eyeball movement and pupillary response. The facial nerve controls the forehead and eye closure.

15) ANSWER: b) administer high-flow oxygen
This patient is experiencing an allergic reaction. Following the priorities of patient assessment and care (Airway, Breathing, Circulation...), the priority intervention is to administer high-flow oxygen (under the breathing priority step). Obtaining IV access would come under circulation. Obtaining vital signs would be in the secondary assessment. Administering epinephrine 0.3 mg would be indicated, after the primary assessment is completed, however.

16) ANSWER: a) Cushing's syndrome
The symptoms being described are typical of Cushing's syndrome. This is a condition of increased circulating glucocorticoids and corticotropin. Addison's disease is due to a decrease in cortisol and is manifested by hyperpigmentation, dehydration, hyponatremia and changes in sexual characteristics. SIADH is a result of an excess of antidiuretic hormone, causing a dilutional hyponatremia. Grave's disease is a manifestation of hyperthyroidism and common symptoms are exophthalmos (bulging eyes) and goiter.

17) ANSWER: c) 300-499 mg/dl
At a BAL of 100-249 common symptoms are slurred speech, ataxia and slowed reaction time. 250-299 is more likely associated with a degree of sedation, significant incoordination, a depressed mental state and nausea and/or vomiting. 300-499 BAL is likely to incur impaired deep tendon reflexes, comatose state, seizures, nystagmus and possible hypoventilation. A BAL over 499 more may cause cardiac and/or respiratory arrest.

18) ANSWER: d) hypocalcemia
Chvostek's sign is seen when the clinician taps the patient's facial nerve, just below the temple. A positive result is observed when the facial muscles twitch. While not 100% sensitive, it is potentially associated with a calcium deficit.

19) ANSWER: b) hyperkalemia
During heatstroke there may be massive amounts of muscle tissue breakdown. In addition to myoglobin being released and possibly damaging the renal apparatus, potassium is released as well causing an increase in the serum potassium levels. Magnesium is not typically a concern for the heat stroke patient. Due to the resultant anaerobic metabolism of the body's environment, a metabolic acidosis may ensue as well. In response to this a compensatory respiratory alkalosis may be seen, not a respiratory acidosis.

20) ANSWER: b) pericardiocentesis
This patient is exhibiting findings of cardiac tamponade. The symptoms of JVD, muffled heart sounds and hypotension are called Beck's Triad and associated with cardiac tamponade. Pericardiocentesis is performed to relieve the expanding pericardal sac which is compromising cardiac output. Needle decompression would be indicated for tension pneumothax. This is evidenced by absent or markedly diminished/decreased breath sounds, JVD, tracheal deviation and hypotension. Peritoneal lavage is done as a diagnostic assessment to look for the presence of intra-abdominal bleeding. A cricothyrotomy would be done as an alternate emergency airway when traditional means of securing the airway were not successful or available.

21) ANSWER: c) administer positive pressure ventilation
According to the NRP guidelines, the next intervention to perform based on the information given is to provide positive pressure ventilation. This is term neonate who is NOT breathing and is limp. Based on this initial assessment, the next steps are to: warm the baby, clear the airway if indicated and to stimulate. The question tells you that the airway is clear so suctioning is NOT needed at this time. As such administering positive pressure ventilation is the next step to be done. Performing an APGAR score is an assessment not an intervention. Establishing IV/IO access is not the priority for this patient. If there was an indication that the airway was compromised or not clear, then suctioning with the bulb suction would have been the next step.

22) ANSWER: d) digoxin toxicity
The classic symptom associated with digoxin toxicity is the patient will report "yellow halos" in the visual fields. Cyanide toxicity is often associated with the smell of "bitter almonds". Lithium toxicity common symptoms are polyuria, tremors and diarrhea. Iron toxicity is often manifested in gastrointestinal symptoms (pain and bleeding).

23) ANSWER: b) bacterial infection
Causes of priapism include: sickle cell disease, use of anticoagulants, leukemia, spinal cord injury, psychotropic drugs, and a penile or urethral tumor. Bacterial infection is not considered a likely cause of priapism.

24) ANSWER: d) lower arm
Compartment syndrome is most likely to occur in the lower arm, hand, lower leg and foot. These areas have limited ability to expand when tissue compartment pressures increase.

25) ANSWER: a) long spine board
A long spine board may be considered as an initial immobilization device. A pneumatic anti-shock garment (trousers)/PASG, while they may be used and considered, are not an initial stabilization device. A traction splint is indicated for femur fractures, not pelvic fractures. Long leg splints aren't indicated for pelvic fractures.

26) ANSWER: b) femur fracture
The most common fracture pattern for adult pedestrian versus motor vehicle is a femur fracture. The other fracture types are possible, but they aren't the most common.

27) ANSWER: b) myoglobinuria
Myoglobin may be found in the urine as a result of the muscle tissue breakdown associated with compartment syndrome, or from the associated traumatic injury which preceded the compartment syndrome. In anticipation of the possibility of myoglobinuria, a minimum urine output of 75-100 ml/hr should be maintained. DIC and anemia aren't likely complications of compartment syndrome. Osteomyelitis would be a concern if there were open bone fractures into which infection could transmit, but by itself, it is not a likely complication of compartment syndrome.

28) ANSWER: d) greater than 12 hours
Fat embolism syndrome will be evident most likely between 12 to 72 hours following a traumatic event.

29) ANSWER: b) urethral injury
Urethral and bladder injuries are commonly associated with pelvic fractures. Muscle spasms are more likely to be associated with long bone (femur) fractures. Neurovascular compromise is more likely to be evident in extremity trauma. Rhabdomyolysis may be associated with any musculoskeletal trauma, but it isn't highly specific for pelvic fractures by themselves.

30) ANSWER: b) use words such as dead, died, or death
Using words such as dead, died or death helps to reinforce reality. It also prevents denial and will add support to the overall grieving process. Telling family members "everything will be all right" or "it was for the best" will minimize their (the family's) feelings. Telling the family the patient didn't suffer may be unrealistic, and should only be stated if known to be absolutely true.

31) ANSWER: a) The bite may go unnoticed due to minimal pain.
Typically at the moment of the bite it is no more painful than a pinprick and may go unnoticed at that time. The pain may progress over the next few minutes however. The venom of the black widow causes a cholinergic excess (similar to organophosphate poisoning), and the treatment would be an anticholinergic (such as atropine). It is the female of the species which tends to be aggressive and bite, the males are not usually venomous. The most common symptom of envenomation reported is abdominal cramping.

32) ANSWER: c) to determine if the patient's condition is life threatening
The priority in any patient scenario is to assess for any life threatening conditions which must be assessed first. The other assessments (history of prior suicidal events, presence of support systems and true intent of self harm) will be assessed for after assessing for any life threats.

33) ANSWER: c) All bat injuries/wounds must receive rabies vaccine.
Bats are amongst, if not the most contagious animals to carry rabies. Current infectious disease guidelines recommend rabies vaccination prophylaxis to any and all bat injuries/bites regardless of the circumstances. When injecting the rabies toxoid, it is given in intra-muscular sites. It is the rabies immune globulin that is injected around the wound. Rabies immune globulin is a passive vaccination which starts working immediately. It does not provide long term protection. Rabies toxoid is a type of active immunization and it requires time for the body to respond and build up antibody stores.

34) ANSWER: d) her husband
Permission for organ and tissue donation must be obtained from the next of kin in the following order: Spouse, adult son or daughter, parent, adult brother or sister. Although the patient is separated from her husband, under law he is still her husband and considered the next of kin. In this situation, the transplant coordinator would want to be sure there was agreement among all parties before donation takes place.

35)   ANSWER: a) 0 to 15 mm Hg
The skull is a rigid, closed box containing the brain and its associated structures, cerebrospinal fluid and blood. Normal ICP in an adult is from 0 to 15 mm Hg. As one component of the skull increases, others must decrease to balance this fine-tuned pressure. When ICP reaches 20 mm Hg or more of pressure, this is called intracranial hypertension. If ICP is not reduced to normal range, ischemia or necrosis of brain tissue may occur, as the oxygenated blood flow becomes restricted.

36)   ANSWER: a) intravenous fluids (IVF) immediately
This patient is in a moderate dehydration state which requires immediate fluid and electrolyte replacement, especially in children who can volume deplete very quickly. Administration of oral replacement fluids would be acceptable in a mild dehydration state, but will not prevent complications at this point. Prochlorperazine (Compazine®) is not usually administered in children because of the side effects. Antibiotics are used for bacterial gastroenteritis, especially traveler's diarrhea. However, most gastroenteritis is viral and must run its course without the use of antibiotics.

37)   ANSWER: a) heart rate 104 bpm
The signs and symptoms of a pulmonary embolus may be highly non-specific, however there are some that have been commonly associated with a PE. Pleuritic chest wall pain is a symptom, not a sign. Room air SpO2 and hemoptysis have not been validated as markers for pulmonary embolus. Tachycardia (> 100 bpm) is associated with approximately 44% of pulmonary embolus cases. Other signs frequently encountered with a PE include: tachypnea > 16/min (96%), rales (58%), accentuated 2nd heart sound (53%), fever > 100.4 (43%), diaphoresis (36%), S3 or S4 gallop (34%).

CEN Q & A HANDBOOK VOL IV

38) ANSWER: d) conduct a one-on-one discussion with the patient
Preprinted instructions may not be useful in an elderly patient due to reduced visual acuity. Handwritten instructions may not be useful due to the patient's visual acuity. Instructing the patient to call his physician with questions is not appropriate. The nurse is responsible for teaching the patient wound care. Having a one-on-one discussion with the patient enables the nurse to discuss pertinent information on wound care as well as demonstrate appropriate dressing application.

39) ANSWER: c) a respiratory rate of 11/min
Symptoms of magnesium toxicity are respiratory depression and loss of deep tendon reflexes. Magnesium sulfate will not affect blood pressure, heart rate or temperature.

40) ANSWER: b) acute bronchitis
For COPD, the chest x-ray would show hyperinflated lungs and a flattened diaphragm. For asthma, the symptoms would include increased mucous secretion and reversibility of the bronchoconstriction. If the patient had pneumonia, the chest x-ray would indicate infiltrates being present.

41) ANSWER: d) suction airway, assist ventilation with BVM, start 2 large bore IVS, attach cardiac monitor
Following the priorities of assessment/intervention, the sequence is A (airway), B (breathing), C (circulation), and D (disability/neuro) followed by the secondary survey. The only answer which follows those steps in order is "D".

42) ANSWER: a) perform a jaw-thrust and assess the airway
All patient assessments begin with assessing the effectiveness of airway/breathing. Granted, in a possible CPR scenario you simply assess for "absent, abnormal, or gasping" breathing, this is a trauma presentation. In this case the standard PHTLS or TNCC patient assessment sequence of "ABCD" (primary survey) is indicated. In this case (due to the mechanism of injury) one would perform a jaw-thrust to open and visually inspect and assess the airway. A head-tilt might be considered if the jaw thrust was ineffective. Checking for a pulse would come under step "C"-circulation. Assessing breathing and starting PPV (if indicated) would come under step "B"-breathing.

43) ANSWER: c) the readiness to leave the perpetrator
Patient safety is always paramount. However, providing safety for the patient in this situation can only be done if the patient is ready and willing to accept it. If the patient is ready and desires to leave the perpetrator, emergency services will assist the patient to a safe place and appropriate community resources. The concurrent use of drugs and/or alcohol may play a role but is not the priority assessment based on the information given. Questioning the patient about remaining in an abusive relationship causes undue defensiveness and attempts at validation by the patient, neither of which are warranted. A history of prior victimizations may be useful to assess the patient's predisposition towards these relationships, but it does not serve the patient's need for immediate safety if they are willing to accept.

44) ANSWER: c) 1 week - 1 year
SIDS is most common during the age range of 1 week - 1 year with the peak being between 2-4 months.

45) ANSWER: c) The patient is hypoxic
Pulse oximetry measures how much the hemoglobin carrier is saturated with oxygen. Oxygenation is a measurement/parameter of the tissues to receive oxygen. Ventilation is the ability of the lungs to exchange gas across the alveolar membrane. Pulse oximetry measures neither of these and any statment about ventilation/oxygenation using pulse oximetry would not be accurate. Hypercarbia (an elevated CO2 level) is not detectable by pulse oximetry, although one might assume that hypercarbia is present in the patient who is not ventilating well. However the question does not tell us that and we cannot assume that it the problem. Hypoxia, however, can be reasonably assumed from the information given. If the oxygen saturation (SpO2) is low, we can deduce that there is an absence of oxygen being carried to the tissues. Granted things like the accuracy of the reading, anemia, perfusion to the digit being measured etc. all may influence the pulse oximetry reading, we are not to assume these things as the question does not convey this. Therefore the best answer to the question is that hypoxia must be assumed to be present.

46) ANSWER: b) hypocalcemia and hypothermia
PRBCs have citrate in them which acts as an anticoagulant to preserve the product life. Multiple PRBC transfusions may exceed the liver's normal process to clear the citrate. As a result the high levels of citrate in the body will bind with ionized calcium rending it useless and as such, causing symptoms of hypocalcemia. The risk of hypothermia also exists due to massive amounts of possibly less-than-body-temperature fluids.

47) ANSWER: d) ventricular tachycardia
Supraventricular tachycardia is a narrow complex tachycardia. An idioventricular rhythm is bradycardic. Atrial fibrillation is a narrow complex rhythm and has an irregular rate. Ventricular tachycardia is a wide complex rhythm and the most likely one to cause the patient to experience sudden unresponsiveness.

48) ANSWER: c) succinylcholine
This patient is experiencing a syndrome called "wooden chest" or "chest wall rigidity". It is a possible complication of several opiates given in high dosages and at rapid rates. The mechanism is not well defined. Fentanyl is the most commonly cited drug for causing this. It causes a rigid (tight) musculature of the skeletal muscles, specifically of concern the thoracic and abdominal areas. This prevents the patient from being able to effectively and spontaneously breathe. As well, the health care provider may find it difficult or impossible to ventilate the patient due to the rigidity. In the patient scenario given, the priority is to establish a secure airway. This is best accomplished most reliably and quickly with the administration of a neuromuscular blocking agent (NMBA) such as succinylcholine. Subsequently, the patient will need to be intubated and mechanically ventilated. If the chest wall rigidity were to occur in a post-extubation period (for example in PACU), then a rapid trial of narcan might be considered to avoid having to re-intubate the patient. Narcan administration may however, cause some withdrawal type effects and a surge in the dopamine neurotransmitter. Romazicon is the antidote for a benzodiazepine overdose/toxidrome. Lidocaine would not be indicated for this patient at this time.

49) ANSWER: b) glaucoma

Contraindications to the use of Anectine (succinylcholine) include: malignant hyperthermia, penetrating eye injury, glaucoma, increased intracranial pressure and those conditions which may precipitate hyperkalemia (48-72 hours post burn-crush injury, neuromuscular disorders [MS, Myasthenia, ALS]). Hypernatremia, concussion, and reactive airway disease would not be factors.

50) ANSWER: a) digoxin toxicity

Elective cardioversion should be avoided in the patient who is digoxin toxic. There is an increased likelihood of inducing an irreversible ventricular tachycardia. Hypermagnesemia, cyanide toxicity and methemoglobinemia are not contraindications to cardioversion.

51) ANSWER: d) a cat bite wound

Cat bites are typically puncture type wounds. Puncture wounds by themselves carry a significantly higher chance of becoming infected than most other wounds. Additionally, the bacterium pasteurella m. is very likely to cause an infection and requires specific antibiotics (augmentin is usually preferred). Scalp lacerations typically don't become infected. Acute sinusitis should be treated with conservative measures (over-the-counter medications) first. Antibiotics usually aren't used for the first week or so. Strep pharyngitis will clear on it's own. Antibiotics are given to prevent rheumatic heart disease in those people at risk. Of all the wounds mentioned, the cat bite (puncture) is the most likely to become infected and mandates antibiotics.

52) ANSWER: c) isoproterenol
In the heart transplant patient there is no intact vagal innervation to the heart. Therefore, atropine would be ineffective. Adenosine is used for narrow complex tachycardia (SVT). Both isoproterenol and epinephrine would have beta 1 activity on the heart, but isoproterenol would be more effective and indicated as a first line drug for symptomatic bradycardia in the heart transplant patient.

53) ANSWER: d) hypotension
Pediatric patients have an ability to compensate for hypovolemia by increasing the intrinsic heart rate and peripheral vascular resistance. In doing so, up to one-third of the blood volume may be lost before a drop in the blood pressure ensues. Tachycardia would be an early sign according to his mechanism. Decreased bilateral breath sounds would more likely be associated with a pneumothorax in the setting of trauma. Diaphoresis would be an early sign due to the vasoconstriction during compensation.

54) ANSWER: b) phentolamine (Regitine)
When dopamine extravasates into the tissues, it has a profound adrenergic like effect, causing profound tissue vasoconstriction and subsequent ischemia and possible necrosis. Phentolamine (Regitine) injected around the area of the extravasation acts to reverse these effects and potentially salvage the affected tissue by encouraging perfusion. Phentolamine (Regitine) is a catecholamine blocker. Calcium gluconate topically is the treatment for hydrofluoric acid topical exposures. Sodium thiosulfate is one of the three components of the cyanide antidote kit. Hydralazine (apresoline) is a vasodilator used to lower blood pressure.

CEN Q & A HANDBOOK VOL IV

55) ANSWER: a) hypotension due to a tachyarrhythmia
Dopamine has strong alpha, beta and dopaminergic receptor agonist properties. Tachycardia is a potential side effect of dopamine due to these effects, and as such, contraindicated in tachyarrhythmia induced hypotension. Dopamine is indicated for treatment of normo-volemic hypotension and hypovolemic hypotension following failed fluid resuscitation. Dopamine can be used with dobutamine to increase systemic blood pressure and cardiac output.

56) ANSWER: b) 45 ml/hr
   First change 1gm to 1000 mg.
   Next divide 1,000 mg by 250 ml.
   That yields a concentration of 4mg/ml.
   Now... 3mg/min x 60 = 180
...and 180/4 = 45 ml

57) ANSWER: B) 14 ml/hr
First calculate the concentration of the dobutamine. In this case it is 500 mg divided by 250 ml which equals 2 mg/ml (2,000 mcg/ml). Next consider the rate. It is ordered at 6mcg/kg/min for this patient.
   Using our IV drip rate formula:
   [(mcg/kg/min) x (pt kg) x 60 min]/concentration =
   [(6mcg/kg/min) x (75 kg) x (60 min) = 27,000
(27,000)/(2,000 mcg/ml) = 13.5, rounded up to 14 ml/hr

58) ANSWER: d) acts to depress the central nervous system (CNS)
Magnesium sulfate depresses smooth, cardiac and skeletal muscle tissues as well as the CNS. It is used to treat certain ventricular arrhythmias, torsades de pointes, and hypomagnesemia. It is also used for pre-eclampsia and eclampsia both for the effect to depress the CNS as well as the effects on the skeletal muscle tissue. Magnesium sulfate would be contraindicated in renal insufficiency and atrioventricular heart block.

59) ANSWER: a) it should be diluted in D5W and in a glass bottle
Nitroglycerin must be mixed in D5W and kept in a glass bottle. Additionally, nitroglycerin is administered via non filtered, non-PVC tubing. A plastic container or tubing will absorb approximately 80% of the nitroglycerin drug molecule. An opaque wrapper is used for an IV of nitroprusside (Nipride).

60) ANSWER: b) physostigmine
Anticholinergic crisis is characterized by hypertension, tachycardia, mydriasis, decreased bowel sounds, urine retention and dry skin. Also use the memory phrase: RED as a beet (cutaneous vasodilation), DRY as a bone (anhidrosis), HOT as a hare (anhydrotic hyperthermia), BLIND as a bat (non-reactive mydriasis), MAD as a hatter (delirium, hallucinations), FULL as a flask (urinary retention). Physostigmine is an anticholinesterase inhibitor, which potentiates the amount of cholinesterase available at the neuron junction, thereby reducing symptoms. Naloxone is the treatment for opiate overdose. Dopamine and dobutamine are used to increase cardiac output and systemic blood pressure.

61)   ANSWER: b) chest pain
The concern for the patient having chest pain is the possibility of an acute myocardial infarction. They must be triaged immediately and an acute MI either ruled out or considered. If an acute MI is present, getting the patient to an interventional cardiac catheterization lab is paramount. Severe abdominal pain, seizures and congestive heart failure would all warrant prompt attention and evaluation, but within a 30-60 minute time period, unless another mitigating factor were present (hypotension, status epilepticus, or hypoxia).

62)   ANSWER: c) fever, depressed level of consciousness and altered liver function tests
Reye's syndrome is associated with liver damage incurred when aspirin has been taken during a viral illness. The child will present with fever and possible depressed level of consciousness. Laboratory findings will reveal damage to the liver and alterations in markers of liver function (bilirubin, protein/albumin, PT/INR, ammonia). Joint pain, red macular rash and low grade fever, indicates most likely rheumatic fever. Peripheral edema and fever > 5 days with a strawberry tongue suggests Kawasaki's disease. A red, raised, bull's eye rash, malaise and joint pain suggests Lyme disease.

63) ANSWER: c) decreased PaO2
Hypoxemia is a universal finding in ARDS regardless of the cause. Initially, due to hyperventilation, the PaCO2 will be low and then later it will rise due to respiratory fatigue. The HCO3 may be low due to decreased tissue oxygenation. The reduced oxygenation leads to anaerobic metabolism and rising lactate levels. The HCO3 combines with the lactate thereby reducing the circulating HCO3 amount. Carboxyhemoglobin may be elevated in the patient with an inhalational lung/pulmonary injury and that may, in turn, progress to ARDS. However, there are other causes besides inhalational injury that can lead to ARDS. An elevated COHb isn't specific to ARDS.

64) ANSWER: a) Solumedrol IV
This patient is experiencing a potential ARDS (acute respiratory distress syndrome) presentation. ARDS is typically triggered by a significant insult to the lung parenchyma. The insult/damage causes an overwhelming inflammatory response of the lung tissue with increased permeability of the alveoli that allows reactive/inflammatory fluid to rush to the damaged tissues. Because of the increased permeability, the fluid overload essentially fills the lungs with fluid quickly, decreasing the vital capacity and effective oxygen exchange surface area. In essence the patient fills their lungs with fluid and drowns them. In this case the best treatment of those listed is the solumedrol, which will address the overwhelming inflammatory response of the lung tissue. Magnesium sulfate does not have a role here. Although it is a bronchodilator, it is not as expedient as an inhaled beta-2 agonist (proventil, albuterol etc). Thiamine is not relevant here. Racemic epinephrine is contraindicated due to the patient's tachycardia. An inhaled beta-2 agonist would be preferred.

65) ANSWER: b) pulmonary angiography
Pulmonary angiography will confirm the presence of a pulmonary embolus when one is present. V/Q scan will not actually confirm the presence, rather it shows where there is a ventilation/perfusion mismatch in the area of question of the lung tissue, and from that information we assume a pulmonary embolus is present. A bronchoscopy would be used to gain direct access to the pulmonary tree for tissue or secretion sampling. A chest x ray would give very non-specific findings.

66) ANSWER: a) 26 year old female who smokes cigarettes
The D-dimer test is useful to rule out a pulmonary embolus (PE). There are many conditions which might cause it to elevate, they are non-specific. Therefore, the goal is to identify the patient who is low-risk or who has a low pre-test probability and use it for that person. The clinician desires the D-dimer to come back negative and then the PE workup can cease. Using tools such as the PERC (Pulmonary Embolism Rule-out Criteria) will help identify the appropriate patients. In this case, patient's B, C, D all have evidence based findings which make the D-dimer testing not useful. Heart rate >100, swollen lower extremity and exogenous estrogen are all evidence based risk factors applicable to the PERC tool. And the presence of them along with the differential diagnosis of PE, negates the use of a d-dimer test, and warrants more definitive testing (such as CT Angio). Cigarette smoking is a risk factor for thrombo-embolism, however, the presence of that alone does not trigger the PERC decision tool and with only cigarette smoking being a risk, this patient would likely have a negative D-dimer and could have a PE excluded.

67) ANSWER: a) thumbprint sign
Given this child's history and assessment, the presumptive diagnosis must include epiglottitis (sudden onset, temp > 103, drooling, stridor, tripod position). The presence of the thumbprint sign (or thumb sign) on lateral neck x-ray is seen often in epiglottitis. The thickened, possibly abscessed epiglottis may be seen with this technique. Steeple sign is associated with croup (laryngotracheobronchitis). Subglottic narrowing is associated with bacterial tracheitis. Supraglottic narrowing isn't a diagnostic indicator.

68) ANSWER: c) administer a 1 liter colloid bolus IV
This patient is in neurogenic shock as evidenced by the low heart rate, low blood pressure and the difference in skin assessment above and below the level of the injury. Neurogenic shock is one the types of distributive shock (the others: anaphylactic and septic), resulting in a profound peripheral vasodilation. The initial treatment for this hypotensive patient would be a fluid bolus/challenge. If the patient failed to respond to this challenge, then an IV of dopamine or norepinephrine (Levophed) may be considered. Atropine would only elevate the heart rate and do nothing to address the vasodilation. Cardiac pacing would have the same effect as atropine.

69) ANSWER: c) aortic rupture
All of the answer choices are potentially associated with a first or second rib injury. The most serious however, would be an aortic rupture, in which case death would be almost instantaneous. Possible symptoms of an aortic rupture include: neuro/vascular/circulatory deficits in the lower extremities (often associated with concurrent spinal cord damage). Also unexplained hypotension and chest or back pain.

70) ANSWER: a) bilateral crackles in the lung bases on auscultation
The nature of the pulmonary contusion is that there is an inflammatory response that occurs in the lung tissue and subsequently fluid may accumulate in the lungs due to the increased permeability. This fluid may accumulate to a true fluid overload, thereby decreasing vital capacity and the ability to oxygenate. An increase in temperature and WBC is an expected finding due to the inflammatory response, however, these are not associated with the highest predictive value for poor outcomes. Hemoptysis may also occur with a pulmonary contusion, however, by itself, it is not a poor predictor for patient outcomes.

71) ANSWER: a) tetanus toxoid is an active vaccination
Tetanus toxoid is given and causes the body's immune system to develop it's own antibodies to the tetanus pathogen. This is an active process of the immune system (as compared to a passive vaccination with the use of immune globulins). Pregnancy is not an absolute contraindication, there are certain criteria for when tetanus vaccine should be administered during pregnancy. HIV is not a contraindication, in fact, one with HIV should receive tetanus vaccination to confer protection against this disease. Tetanus immune globulin (Ig) is a passive form of vaccination. As it does not trigger the body's immune system to manufacture antibodies, there is no long term protection conferred.

72) ANSWER: a) the presence of currant jelly stool
Bowel intussusception occurs when a portion of the bowel "telescopes" into itself. Where the telescoping has occurred, the tissue is inflamed and secretes mucous containing fluid. Additionally, the richly vascular lining is likely to have capillary ruptures and bleeding also. The combination of the mucous and the fresh blood mix to create stool that appears as currant jelly. The term "currant jelly" is pathognomonic with bowel intussusception. Diarrhea for 24 hours is non specific for any one particular pathology. While an elevated temperature is a possibility with bowel intussusception, it is not specific. There are several conditions which may/may not cause a fever of that magnitude. Projectile vomiting is nearly pathognomonic for pyloric stenosis.

73) ANSWER: a) motor vehicle collisions
Motor vehicle collisions account for 60-80% of all chest trauma-related deaths. Falls, assaults and firearms are responsible for the remaining 20-40%.

74) ANSWER: c) decreased ventilatory drive
A COPD patient has a respiratory drive regulated by the amount of PaO2 versus PaCO2 (the healthy patient). This is called a hypoxic drive. A decrease in PaO2 becomes a stimulus to breath and an increase becomes the stimulus to not breathe. If excessive supplemental O2 is administered the patient will lose the hypoxic respiratory drive and respirations will slow or even stop. As PaO2 increases, and respirations decrease and the PaCO2 will increase, not decrease. COPD has a ventilation/perfusion mismatch by it's very nature. The COPD patient's V/Q mismatch is due to the chronic overdistention of the alveolar walls causing less functional surface area for the exchange of gas. The increased O2 levels for this patient would not affect the vital capacity.

75) ANSWER: b) routine ACLS/resuscitation guidelines will be followed
During a hypothermia resuscitation, normal resuscitation guidelines and ACLS protocols will be followed until a core temp of 86F is achieved. Below 86F, normal physiological processes and pharmacological responses are not predictable. Once 86F is achieved, then physiological processes/responses are more predictable. In hypothermia, pt will more likely be in an acidotic state due to the anaerobic metabolism and lactic acidosis. Shivering is diminished or absent (more likely) below 86F. In hypothermia, the pupils are more likely to be dilated.

76) ANSWER: b) paroxysmal
Pertussis follows a predictable clinical course of progression. During the initial catarrhal stage, the patient's symptoms are very non specific and are consistent with a generic upper respiratory infection: rhinorrhea, low grade fever, mild cough (not whooping cough yet), malaise. The patient is contagious during this time. This is followed by the paroxysmal stage in which the distinctive "whooping" cough is observed. The patient may experience several successive bouts of coughing paroxysms, sometimes to the point of post-tussive emesis. Typically the clinician makes the diagnosis at this time. The convalescent stage is the subsequent 2-3 weeks during the recovery phase. There is an increased susceptibility to respiratory infections during this time. Coryzal is not a stage of pertussis.

77) ANSWER: d) tracheal shift and jugular venous distention
For this patient with an open pneumothorax, the worst case scenario would be progression to a tension pneumothorax. An oxygen saturation of 93% in this scenario is adequate at this time. A blood pressure of 110/70, albeit lower than normal, is adequate as well at this time. Decreased breath sounds may be a worsening, but by themselves, do not indicate the presence of a tension pneumothorax. The presence of tracheal shift/deviation and jugular venous distention are consistent with a tension pneumothorax, and indicates the immediate need for action, such as a needle chest decompression.

78) ANSWER: b) hypoxia
Hypoxia in the pediatric patient leads to bradycardia, and unless corrected, then to asystole. Treatment of any bradyarrhythmia in the pediatric patient requires assessing and treating any oxygenation issues first. The other options listed may contribute to cardiac arrest in the pediatric patient, but are much less likely or common.

79) ANSWER: b) gallbladder disease and alcohol abuse
When considering all causes of acute pancreatitis, biliary tract disease and alcohol account for over 80% of the cases. All the others are possible causes, however they are all possible causes.

CEN Q & A HANDBOOK VOL IV

80) ANSWER: d) manage the airway
Alkali ingestion injuries cause corrosive burns of the upper GI tract. Due to the ingestion, there is an increased chance of aspiration which may lead to esophageal perforation of laryngeal/tracheal edema and occlusion. The airway must be managed aggressively in this case. Vomiting should not be induced as it may cause the alkali substance to return up the GI tract causing further damage. Due to the risk of esophageal tissue breakdown and perforation, insertion of a nasogastric tube should be avoided, except unless truly necessary and then only under fluoroscopy or endoscopy. Steroids would not be beneficial in this scenario.

81) ANSWER: d) the patient is currently taking dilantin (phenytoin)
Isoniazid is contraindicated with concurrent dilantin therapy as it may decrease the excretion of dilantin and enhance it's effects possibly to toxic levels. If isoniazid is mandatory, the patient's dilantin dosing should be adjusted to reflect this. The use of lasix, the presence of glaucoma or coronary artery disease have no documented concerns for concurrent administration of isoniazid therapy.

82) ANSWER: c) ages 31-60
A pheochromocytoma is a neoplasm affecting the adrenal glands, causing a hyperfunctioning with a surplus of circulating catecholamines. Common symptoms are sustained hypertension, visual disturbances, headaches, hyperglycemia and excessive perspiration. While it can be diagnosed in any age group, the most common group is between 31-60 years old when first identified. Seldom/rarely does it occur in the patient over 65 years old.

83) ANSWER: b) bloody diarrhea
The findings of a mesenteric infarction, especially early on are very non-specific. One of the hallmark findings is "disproportionate pain" or "pain out of proportion" to the abdominal exam. However given the choices listed above, bloody diarrhea is the most specific finding associated with this condition. The vomiting and diarrhea are non specific. There is no direct relationship between mesenteric infarction and free air under the diaphragm. A perforated bowel/viscous is most often the culprit with free air under the diaphragm.

84) ANSWER: d) administering fluids
Primary treatment will be to replace fluids. The specific choice of fluid will be determined by the cause of the imbalance. A serum osmolality should be considered to assist with this identification. If it is determined the patient is hypovolemic, then replacement would be started with normal saline and then titrated down to a hypotonic solution. Giving Kayexalate will increase the serum sodium level, worsening the problem. Likewise diuretics are sometimes implicated in contributing to the problem and would make things worse. Activated charcoal would be ineffective as this is not a toxidrome for which absorbing a toxin/molecule would be a solution.

85) ANSWER: a) sickle cell crisis
Sickle cell crisis in children typically will manifest as pain in the hands or feet joints. There is usually a degree of swelling present also. Meningitis isn't typified by joint swelling and there is no indication in this patient of CNS involvement, nor nuchal rigidity or meningeal irritation. Hepatitis would be characterized by jaundice, clay-colored stool, and concentrated urine. Hand swelling isn't consistent with hepatitis. Discoid lupus affects the skin/connective tissues. It's classic findings are malar erythema (butterfly rash), joint pains, fever and behavioral changes.

86) ANSWER: d) procardia
This patient is experiencing autonomic dysreflexia (AD) which can be a potentially life threatening emergency. AD occurs in quadriplegics and those with higher spinal cord injuries usually of the T6 level or higher. AD results in a hypertensive emergency and the priority is to lower the blood pressure urgently. Of the agents listed, procardia would be the agent to use to lower the blood pressure. Verapamil, while it may lower the blood pressure some, is primarily used for treatment of SVT for rate control. Demerol would help with the headache, but not with the root cause of the headache, namely the blood pressure. Heparin would not be used as it may increase the risk of a cerebral or subdural bleed secondary to the accelerated hypertension.

87) ANSWER: b) a urinalysis showing bacteriuria
Epididymitis is typically the result of an infection either bacterial or an STD (sexually transmitted disease). This causes swelling/enlargement of the epididymis and associated scrotal pain. Due to the infectious nature, the CBC would show an elevated WBC count (leukocytosis) and not leukopenia (a decreased WBC count). An elevated creatinine level indicates a renal problem not a scrotal/testicular problem. If an ultrasound was performed, epididymitis would show increased blood flow (hyperperfusion), not hypoperfusion.

88) ANSWER: a) temporal lobe
People with receptive aphasia are unable to understand language in its written or spoken form. Even though they can speak with normal grammar, syntax, rate, and intonation, they cannot express themselves meaningfully using language. The temporal lobe contains the auditory association area. Damage to this lobe in the dominant side causes the patient to hear words, but not understand their meaning. When the parietal lobe is affected, the patient's ability to identify special relationships with the environment. Damage to the occipital lobe affects the visual sensory processing. In this case the patient can see objects, but cannot identify them. Damage to the frontal lobe manifests as personality and memory changes/alterations.

89) ANSWER: b) proteinuria, headaches, double vision
Pregnancy induced hypertension includes: pregnancy, a blood pressure of 140/90 or greater, OR a SBP > 15mmHg from baseline or a DBP >10mmHg from baseline, plus some evidence of end organ damage or effect such as edema or proteinuria. Headaches and double vision are common as well. Proteinuria may be a late finding however. Uterine contractions and vaginal bleeding are not typical of PIH. By definition, PIH will not be diagnosed until after 20 weeks gestation.

90) ANSWER: a) seizure activity
This patient is having symptoms of pregnancy-induced hypertension (PIH). There the paramount concern of seizure activity due to the irritability of the CNS. A seizure could potentially compromise the fetal circulation/oxygenation. Cardiac arrhythmias, vaginal bleeding and a precipitous delivery are not complications of PIH.

CEN Q & A HANDBOOK VOL IV

91) ANSWER: c) administer prednisolone and Tropicamide (Mydriacyl)
Iritis does need opthamalogical follow up, but not emergently. Topical (opthamalogical) decongestants would be used for an allergic eye disorder. Opthamalogical erythromycin would be considered for eye infections or potential infections such as conjunctivitis or corneal abrasions. Iritis is an inflammatory problem, as such the use of an opthamalogical steroid is appropriate. Tropicamide is an eye dilator. This would be an appropriate use for iritis. This would help to allevaite the discomfort associated with the affected ciliary muscles.

92) ANSWER: d) The patient should be placed on immediate bed rest.
Immediate bed rest is necessary to prevent/limit further injury. Patching both eyes may be considered but it isn't the priority here. Early referral to an ophthalmologist is appropriate also, but the priority is to limit further damage. Surgical reattachment is appropriate, however, in the interim the patient should be on bed rest.

93) ANSWER: a) administration of high-flow oxygen
A fat embolism is a possible complication of bone fractures. The major concern here is the emboli traveling to the pulmonary vasculature causing possible infarcts or right sided heart failure. The effects are typically systemic. A petechial rash is nearly pathognomonic for this syndrome. The priorities for this patient would begin with airway and breathing, hence administration of high-flow oxygen being the priority. IV fluid bolus (to increase cardiac output) would be indicated depending on the patient's hemodynamic status. Corticosteroids may be considered but are not the priority. Trendelenburg position may be helpful if the patient is hypotensive, however, supplemental oxygen would still be the priority.

94) ANSWER: a) Anticipate the need to x-ray the amputated part before reimplantation.
An x-ray is necessary before re-implantation/re-attachment to rule out fractures and possible foreign material. 60 minutes is not too long to preclude reattachment. Depending on the body part and the circumstances, this time can extend for several hours. The amputated part should be wrapped in sterile gauze and moistened with sterile saline. Then, after a gentle cleansing, it should be placed in a plastic bag, sealed tightly. Then take that plastic bag and place it in another bag with ice in it. The body part should be reassessed intermittently to ensure that it does not freeze.

95) ANSWER: d) the arm
The arm is the most common site for nerve entrapment syndrome. There are different types depending on which nerve pathway is affected. Pronator syndrome involves the median nerve being entrapped at the forearm. Carpal tunnel syndrome is also the median nerve entrapment, but it occurs at the wrist. In cubital syndrome, the ulnar nerve is entrapped at the elbow. The leg, thoracic and vertebral columns do not have an associated entrapment syndrome.

96) ANSWER: d) ankle dislocation, patellar dislocation, hip dislocation
Unrestrained front seat passengers involved in MVCs frequently sustain lower extremity trauma from the knees hitting the dashboard. In this scenario, the common injuries sustained include the ankle/patella/hip dislocations. Other possibilities commonly include: hip fractures, femur fractures. Patella, ankle and foot fractures aren't commonly seen in these settings.

97) ANSWER: a) apply a pressure dressing to the bleeding site
The steps of bleeding control in the Emergency Department setting (which may differ from the pre-hospital or tactical environments) are: direct pressure with elevation, pressure point, then tourniquet.

98) ANSWER: c) diphenhydramine (Benadryl)
Diphenhydramine (Benadryl) 25 to 50 mg IM or IV would quickly begin to reverse the dystonic reaction and symptoms. Prochlorperazine (Compazine) and haloperidol (Haldol) are common agents causing dystonic reactions, not resolving them. Midazolam (Versed) would just make the patient drowsy.

99) ANSWER: c) age of onset in the 40's, legal/financial troubles, multiple divorces
There are 3 main types of bipolar disorder. Bipolar type 1 is the most common form. It is typified by mania, psychosis and a major depressive state. Bipolar II is characterized by recurrent major depressive episodes and hypomanic episodes. Cyclothymic disorder is characterized by: depression that is not enough to meet major depressive criteria and hypomania that does not meet criteria for true mania. From the history, the following are most typical across the board for all bipolar types: multiple divorces, legal/financial troubles and job losses. The age of onset for all types combined occurs from early adulthood through the 40's.

100) ANSWER: d) several 1-1.5 cm diameter circular burns on the posterior thorax
Small circular burns on a child's back are no accident. They may be from healing cigarette burns. Toddlers are very injury prone because of their developmental stage and falls are frequent due to their unsteady gait. Head injuries aren't uncommon and a small area of ecchymosis or a resolving hematoma aren't suspicious in this age group.

101) ANSWER: b) haloperidol (Haldol)
Haldol given via IM or IV route is the drug of choice for acute psychotic behavior requiring chemical constraint. Thorazine may be considered as well, however, it has much more pronounced sedation than Haldol and is possibly associated with more side effects. Lithium is useful for bipolar or manic disorders, but not useful for acute psychosis. Elavil is used for sleep and depression.

102) ANSWER: a) flashbacks, recurring dreams, numbness
Posttraumatic stress can include recurring dreams about the event(s) or vivid flashbacks to the event(s). There may be a general sense of numbness and estrangement from others. Emotional reactions such as denial, anger, guilt, humiliation, fatigue and self-blame are all potentially associated, but not as specifically as the flashbacks, dreams and numbness.

103) ANSWER: b) This patient is exhibiting possible anxiety or panic symptoms.
This patient is demonstrating excessive concern, worry and anxiety about things for about 4 months. There is no logical or factual basis for these. She is also demonstrating signs of excessive autonomic hyperactivity as seen with her rapid heart rate, and difficulty sleeping.

104) ANSWER: b) electrical arcing may occur
The topical medication and patch may serve as an electrical conductor and a defibrillation delivery may allow for the electricity to arc causing injury to the patient or operator(s). The defibrillation will have no effect on the actual medication itself. Reflex tachycardia is not relevant in this scenario.

105) ANSWER: c) severe scoliosis
The severe scoliosis would be restrictive to full chest/lung expansion and as such, would decrease the tidal volume. The decreased vital capacity would cause a retention of CO2 and a decreased ventilatory effort, thus respiratory acidosis. High altitudes would induce a respiratory alkalosis (hyperventilation) as the body attempts to compensate for the decreased partial pressure of oxygen at altitude. Salicylate toxicity induces a metabolic acidosis so the body compensates by forcing a respiratory alkalosis. Likewise with aldosteronism, the ensuing manifestation would be a respiratory alkalosis

106) ANSWER: b) It is defined by the individual.
There is currently no exact and clear definition of critical incident stress. It is a personally defined term and subject to each individual's response and interpretation of the stress or incident. Critical incident stress does not happen that frequently; recalling that critical incident is the operative wording; not just the day to day stress but those overwhelming incidents that have a major/significant impact. Critical incident stress is not just occupation related, it can include on and off scene personnel, and can involve off duty situations as well.

107) ANSWER: a) right coronary
The RCA supplies the inferior and right side of the heart. Leads II, III, and aVf show the inferior portion of the heart. The LAD is shown by the anterior leads V1-V4. The circumflex artery is shown by leads I, aVl, and V5-V6

108) ANSWER: a) 4 ml/hr
To answer this use the following formula:
[(Rx dose mcg/min) * (60 min)]/Concentration (100 mg/250ml= 400 mcg/ml) =
[25 mcg/min * 60 min]/[400 mcg/ml] =
[1500]/[400] = 3.75 ml/hr (rounding up to 4 ml/hr)

109) ANSWER: b) 30 ml/hr
To answer this use the following formula:
[(Rx mcg/kg/min)(kg)(60 min)]/Concentration =
[10 mcg/kg/min)(80)(60)/1600 =
[48000]/1600= 30 ml/hr

110) ANSWER: b) sodium bicarbonate
Sodium bicarbonate stabilizes the ion balance and buffers the pH. This potentiates the effects of catecholamine agents such as dopamine, epinephrine etc. Lidocaine is used for ventricular arrhythmias. Atropine is used for bradyarrhythmias.

111) ANSWER: a) pt is intoxicated, without other findings
Nausea, dizziness and transient confusion are commonly associated with a concussive head injury and not necessarily with a more pathological head injury (subdural, subarachnoid, epidural). The intoxicated patient presents a challenge as often the history is questionable and symptoms may be masked/altered by the effects of the intoxicant. The intoxicated patient having sustained a closed head injury would be the highest priority on this list.

112) ANSWER: a) a fracture of two or more ribs in two places
A flail chest injury is a fracture of two or more ribs in two or more places. This results in a free-floating segment of the chest wall. Paradoxical movement of the chest wall in the area of injury is a common finding. It should be noted that sometimes the paradoxical movement may not be seen until the injured chest wall muscles relax or pain relief is achieved. Flail chest is usually a closed injury and does not typically occur bilaterally.

113) ANSWER: c) a peaked P wave on the EKG
With the presence of a pulmonary embolus (PE) the resultant increase of pulmonary artery hypertension puts increased stress on the heart, specifically the right side. This leads to an increase in right atrial volume and subsequently an altered P wave on the EKG tracing. The ideal lead to observe this is in lead II as here the P wave is taller and more peaked than a normal P wave. A physiologic S2 split sound is a normal finding. Typically, lung sounds are clear in the patient with a PE. A pleural friction rub may be heard with a PE but not a pericardial friction rub.

114) ANSWER: b) correcting the hypoxia with supplemental oxygen
Priorities always follow the ABC's (airway, breathing, circulation). In the case of a pulmonary embolus, oxygenation is going to be affected and should be supplemented with oxygen via an appropriate oxygen delivery device. If the decision is made to administer heparin, it would be loaded with a IV push (bolus) dose first, THEN followed by a continuous infusion. Thrombolytics are only approved for the patient with a confirmed pulmonary embolus that suddenly loses their hemodynamic stability. There may be an associated pain response with a pulmonary embolus, however, administering analgesia, while important, is not the priority. Oxygenation is the highest priority.

115) ANSWER: a) The patient has a fluid volume deficiency.
Normal CVP readings run from 4-10 cm H2O. Above 10 cm H2O indicates a possible fluid volume overload. A reading below 4 cm H2O potentially reflects a fluid volume deficit. Gas exchange or hypoxia is not reflected by CVP readings.

116) ANSWER: a) vasoconstriction
Alpha receptors are predominantly located in the peripheral vasculature. Stimulation causes a vasoconstriction, thereby increasing blood pressure. An example of an agent that does this is Neosynephrine which is used in critical care settings to increase peripheral vasoconstriction and treat hypotension. An alpha receptor antagonist (alpha blocker) would cause vasodilation. Tachycardia is not likely to be seen as heart rate is predominantly regulated by the beta-1 cells in the heart.

117) ANSWER: b) normal or decreasing systolic pressure and rising diastolic pressure
The sympathetic stimulation that occurs in response to a shock state is regulated by specialized cells in the carotids and aorta. A decrease in oxygen is detected with a rise in the carbon dioxide. Catecholamines are released to produce peripheral vasoconstriction and increase the total peripheral resistance. The result of this is an increase in the diastolic pressure as an attempt to increase the preload and thereby the cardiac output.

118) ANSWER: d) bradycardia
As shock progresses cardiovascular dysfunction and decreased cellular functioning leads to decreased cardiac output and inability of compensatory systems to perform. Bradycardia ensues. The sympathetic innervation has very limited compensatory mechanisms. Tachycardia and irritability occur during early shock when the compensatory mechanisms are still intact. 1-2 ml/kg/hour of urine output is normal for a pediatric patient.

119) ANSWER: d) hypovolemic
Dopamine is least useful in hypovolemic shock. For this patient the blood pressure should be raised with circulating volume replacement. Neurogenic and septic shock are characterized by peripheral vasodilation, in which case dopamine would be preferred as it causes peripheral vasoconstriction. The use of dopamine in cardiogenic shock would not be ideal, as it may increase cardiac demand, however it is not deleterious as if it were used in hypovolemic shock.

120) ANSWER: c) ineffective ventilations
The main concern with a toxic methanol exposure is the effects its toxic metabolites have on the brain stem. These metabolites are likely to suppress the brainstem as well as the respiratory control center. The other options are possibilities, but not as much of a priority as the risk for ineffective ventilations.

121) ANSWER: c) tachycardia
Amphetamines are CNS stimulants that cause sympathetic (fight or flight) stimulation. This may include hypertension, tachycardia, vasoconstriction and hyperthermia. Hot, dry skin is seen with anticholinergic toxicities and or jimsonweed. Pupils would be dilated, not constricted in an amphetamine overdose.

122)   ANSWER: a) Immobilize the injured extremity.
The wound of the pit viper is concerning for the localized tissue damage and ensuing compartment syndrome. As such, the wound should be managed along the lines of compartment syndrome. Immobilizing the extremity decreases unnecessary extremity ciruculation which may increase swelling and the compartment pressure. Applying a tight constrcting band would impede circulation and further lead to tissue ischemia. Applying ice would have the same effect, in that it would decrease tissue perfusion, worsening the problem. If the limb was elevated above the level of the heart, this also would decrease tissue perfusion. Immobilization serves this type wound the best to prevent increased pressure while preserving perfusion.

123)   ANSWER: d) the risks and benefits to the patient
The benefits must always outweigh the risks. After assesing the risks, then considerations may turn to the actual mode of transportation and the consent of the patient or family. Insurance status does not play a role in the decision to transfer a patient.

124)   ANSWER: b) cognitive
This is an example of cognitive learning. These skills require thinking and reasoning on the part of the learner in order to integrate the learning. Affective learning involves feelings and attitudes. Psychomotor learning involves the coordination of the brain and the extremities to complete a task.

125)   ANSWER: a) negligence
Negligence is the most common unintentional tort in the health care setting. It is when the medical provider fails to provide the care which is expected and results in injury or death. Malpractice is a more restrictive, specialized kind of negligence, defined as a violation of professional duty to act with reasonable care and in good faith. Assault and battery are intentional torts.

126)   ANSWER: a) statutory law
Statutory law is law made by federal and state legislatures. Medicare law is a federal statute. Ordinances are passed by local jurisdictions such as local parking regulations. Common law is law that is formed by judicial decisions in a courtroom, similar to case precedents. Constitutional law is the supreme law of the land.

127)   ANSWER: c) heart failure symptoms
Due to the lack of a vagal tone/control, the symptoms of tachycardia, chest pain and jaw pain related to a cardiac ischemic/infarction event aren't possible.

128)   ANSWER: c) adrenal crisis
Long term use of corticosteroids causes a suppression of the normal adrena-cortical system. Abruptly stopping the prednisone may precipitate an adrenal crisis. This is manifest as a sudden lack of cortisol. The patient may present in extremis. Psychiatric disrubamces may occur during steroid therapy, but not likely on stopping the steroids. Tardive dyskinesia is associated with patients taking neuroloeptics. Glaucoma is possibly associated with long term steroid use.

129)   ANSWER: b) to reduce cerebral edema
Brain tumors tend to cause localized swelling, edema and inflammation. The use of coticosteroids in this setting is to reduce those effects, and lessen the likliehood of increased intracranial pressure consequences. Absence seizures would be treated with anticonvulsants. Agitation is best treated with anxiolytics or antipsychotics. The choice of an analgesic for this patient would be something non-opiod so as to not mask their neurological status or assessment.

130) ANSWER: b) sodium bicarbonate
Asprin toxicity/overdose leads to a metabolic acidosis. Administration of sodium bicarbonate would help to return the pH to a more neutral state. Additionally, sodium bicarbonate would help to enhance the excretion of the salicylate. Calcium glucontate and magnisium are part of the overdose treatment for hydroflouric acid exposures. Folic acid is part of the treatment for methanol poisoning.

131) ANSWER: d) panic
Panic is the emotion most often associated with critical incident stress (CIS). Hebephrenia (a type of schizophrenia), euphoria and depression are not typically rellated to CIS.

132) ANSWER: b) chest tube drainage
There are two indications for consideration of immediate surgery for this patient with a hemothorax. One is an initial drainage of 1000 ml of blood with chest tube insertion. The other is a continued drainage of 200 ml per hour or more. Both of these indicators warrant surgical intervention to locate and control the source of bleeding. While urine output, CVP and vital signs are important assessment parameters, they do not carry the same importance as monitoring of the chest tube drainage.

CEN Q & A HANDBOOK VOL IV

133) ANSWER: c) sucking chest wound
A sucking chest wound is more likely associated with a pneumothorax or tension pneumothorax. Due the the blunt injury typically involved with a flail chest, pulmonary contusions occur frequently. When a flail chest is present (the fracture of 2 or more adjacent ribs in 2 or more locations), the chest wall demonstrates paradoxical movement of the "free-floating" portion of the chest wall. It moves opposite in direction in comparison to the uninjured area.

134) ANSWER: b) assisting with intubation
This child is exhibiting findings consistent with epiglottitis. Epiglottitis usually progresses rapidly, compared with croup (laryngotracheobronchitis), which usually has an upper respirtory prodrome for a day or two. Epiglottis is also marked by a high fever, drooling and stridor. Typically the child also appears sick or ill, compared with croup, they are generally well appearing. The risk with epiglottis is abrupt airway closure and obstruction. Therefore the child should be stimulated or made anxious as little as possible. Invasive procedures should be avoided in the initial care phase. Even placing a face mask on the child may make them more anxious and increase this risk. Epiglottis should be further assessed and intervened for under a controlled environment whereby the airway can be definitively secured, safely and rapidly.

135) ANSWER: c) Level 3
This patient needs prompt (but not emergent) attention. The patient is a candidate to wait 30-60 minutes for assessment and intervention. This patient will need > 2 resources (labs/xray, IV fluids, antibiotics). Even though they are immune compromised from the chemotherapy, this does not change there resource and severity assessment.

136) ANSWER: a) improved ABG panel
When intubation is performed for a tracheobronchial injury, the inflatable cuff must be passed beyond the tear or disruption. When succesfully placed the patient's oxygenation status will improve as evidenced by improved/improving ABG results. The presence of JVD or a tracheal shift suggest that the endotracheal tube has not been placed correctly and air is escaping via the tracheobronchial rupture causing a tension pneumothorax. A respiratory rate of 36 indicates the patient is not oxygenating effectively.

137) ANSWER: c) serum lactate 6 mmol/L
A patient with septic shock who demonstrates a lactic acid > 4 mmol/L is likely to have a poor outcome. A pH of 7.35 and a CVP of 4 cm H2O are both within physiological parameters. While a BP of 92/56 (MAP = 68), is considered by some to be somewhat hypotensive, it is not the best prognosticator for poor outcomes.

138) ANSWER: b) FFP or platelets
PRBCs contain a minimal amount of clotting factors or platelets. Repeat or multiple units of PRBCs administered may increase the chance of a coagulopathy. Administration of one or two units of FFP/platelets following 5 units of PRBCs is common practice. Albumin is used to expand the vascular volume and isn't related to the administration of PRBCs. Dextran and washed RBCs are not clotting products.

139) ANSWER: b) continue NS infusion at 125 ml/hr
A urine output of 30 ml/hour or more for an adult indicates adequate renal perfusion. Giving another 1 or 2 liters NS bolus isn't indicated at this time. D5W is not indicated in this scenario at this time.

140) ANSWER: a) tachycardia
Cocaine exerts stimulant type effects. Tachycardia is the only symptom listed consistent with a stimulant response. Miosis is constricted pupils – often associated with opiate overdose. In cocaine exposure, one would expect mydriaisis (dilated pupils).

141) ANSWER: c) respiratory exposure
The main concern with an epxosure or ingestion of petroleum distillates is the effect on the respiratory/pulmonary system. The fumes/vapors are heavier than air and can "sink" to occupy the lower/distal airways causing dead space and a resultant risk of hypoxia. Also possible is the risk of a chemical pneumonitis. Typically enough gasoline is not swallowed to cause a problem that exceeds that of the respiratory tree. Cardiac and neurological systems may be affected also, but not as often nor as severely as the respiratory system.

142) ANSWER: c) Rocky Mountain Spotted Fever
Symptoms of RMSF start 2-14 days after infection by a tick, usually in the wilderness/woods. The rash will encompass the soles of the feet, palms, wrists and ankles by the 2-5th day. Eventually it will spread bodywide. Other symtoms of progression may also include delirium, hypotension, and edema. Lyme disease, also a tick borne disease, announces itself with a single "bull's eye" (or target) lesion. It has a halo appearance around it. Poison ivy presents with painful, blistering pustules, usually in a linear fashion. Measles begins on the face and progresses to the torso and then then extremities.

143) ANSWER: c) more patient care space
Ground transport vehicles have larger patient care areas. This may allow for ease of patient care, especially if they are critical or require multiple interventions. Helicopter transport does have the benefit of a faster transport speed and is not affected by road or traffic conditions. The radio communications should be equitable for both ground and air transport.

144) ANSWER: a) affective
Affective learning encompases a change in attitudes, beliefs and values. These changes take time to incorporate and integrate. Cognitive learning is knowledge based. Even if it is highly complex knowledge, it doesn't require the same long term change as with affective learning. Psychomotor learning is associated with physical skills and/or mental coordination, this is best accomplished by practice and repetition. It still does not take as long as affective learning. Social learning does not necessarialy involve a change in attitudes or values, so a statement cannot be conclusively made about the length of time required by social learning.

145) ANSWER: a) quality improvement process
This is a multidisciplinary group, coming together to find a way to improve a patient experience or achieve a better outcome. The purpose is not to generate or to validate a scientific principle or hypothesis. A research process would start with a literature search and generation of a hypothesis which would then go on to be tested. Indicator relevance testing is not part of either research or quality improvement processes.

146) ANSWER: d) self-scheduling
In shared governance models, self-scheduling is most commonly seen. This method emphasizes individual staff accountability and encourages staff involvment. Management scheduling takes the focus away from the staff, and it minimzes the maturity and responsibility of the staff members. A computerized scheduling system is not consistent with the staff being integral and having an input and influence on the scheduling.

147) ANSWER: b) inferior MI
The inferior MI is the more likely of the choices given to cause a pathological bradycardia and AV nodal conduction abnormalities. An anterior MI is more likely to lead to tachyarrhythmias and heart block. Tachycardia is more likely seen with cardiogenic shock and heart failure as the conduction pathways are unaffected, yet the heart is attempting to compensate for a decreased output.

148) ANSWER: a) ruptured aorta
A wide mediastinum is most suggestive of either an aortic anueryism or rupture. In association with the patient's symptoms of chest pain through to the back, aortic pathology must be considered. A ruptured trachea would exhibit a pneumomediastinum on x-ray. A ruptured hemidiaphragm shows free air under the diaphram. A pneumothorax will be visible on x-ray.

149) ANSWER: b) iritis
Iritis is a painful eye condition. It is an inflammation of the ciliary muscles that control pupil dilation, as well as of the uveal tract. When the pupil does react, it is usually sluggish. A central retinal artery occlusion is painless and a loss of vision. Glaucoma typically has a sudden onset of severe pain with a non-reactive pupil. Conjunctivits presents with normal vision, reddened sclera and normal pupil reactivity.

150) ANSWER: c) liver or spleen laceration
A hepatic or splenic rupture/laceration can cause intra-abdominal bleeding. This can put the patient into hypovolemic shock. A diaphragmatic rupture could be present, but no information specific to that is given, such as bowel sounds in the chest or an abnormal chest/abdominal xray. Small bowel contusions typically do not exhibit such markedly abnormal findings on initial examination. An aortic rupture would cause the patient to deompensate and exsanguinate very rapidly.

1) Ⓐ Ⓑ Ⓒ Ⓓ	41) Ⓐ Ⓑ Ⓒ Ⓓ	81) Ⓐ Ⓑ Ⓒ Ⓓ	121) Ⓐ Ⓑ Ⓒ Ⓓ
2) Ⓐ Ⓑ Ⓒ Ⓓ	42) Ⓐ Ⓑ Ⓒ Ⓓ	82) Ⓐ Ⓑ Ⓒ Ⓓ	122) Ⓐ Ⓑ Ⓒ Ⓓ
3) Ⓐ Ⓑ Ⓒ Ⓓ	43) Ⓐ Ⓑ Ⓒ Ⓓ	83) Ⓐ Ⓑ Ⓒ Ⓓ	123) Ⓐ Ⓑ Ⓒ Ⓓ
4) Ⓐ Ⓑ Ⓒ Ⓓ	44) Ⓐ Ⓑ Ⓒ Ⓓ	84) Ⓐ Ⓑ Ⓒ Ⓓ	124) Ⓐ Ⓑ Ⓒ Ⓓ
5) Ⓐ Ⓑ Ⓒ Ⓓ	45) Ⓐ Ⓑ Ⓒ Ⓓ	85) Ⓐ Ⓑ Ⓒ Ⓓ	125) Ⓐ Ⓑ Ⓒ Ⓓ
6) Ⓐ Ⓑ Ⓒ Ⓓ	46) Ⓐ Ⓑ Ⓒ Ⓓ	86) Ⓐ Ⓑ Ⓒ Ⓓ	126) Ⓐ Ⓑ Ⓒ Ⓓ
7) Ⓐ Ⓑ Ⓒ Ⓓ	47) Ⓐ Ⓑ Ⓒ Ⓓ	87) Ⓐ Ⓑ Ⓒ Ⓓ	127) Ⓐ Ⓑ Ⓒ Ⓓ
8) Ⓐ Ⓑ Ⓒ Ⓓ	48) Ⓐ Ⓑ Ⓒ Ⓓ	88) Ⓐ Ⓑ Ⓒ Ⓓ	128) Ⓐ Ⓑ Ⓒ Ⓓ
9) Ⓐ Ⓑ Ⓒ Ⓓ	49) Ⓐ Ⓑ Ⓒ Ⓓ	89) Ⓐ Ⓑ Ⓒ Ⓓ	129) Ⓐ Ⓑ Ⓒ Ⓓ
10) Ⓐ Ⓑ Ⓒ Ⓓ	50) Ⓐ Ⓑ Ⓒ Ⓓ	90) Ⓐ Ⓑ Ⓒ Ⓓ	130) Ⓐ Ⓑ Ⓒ Ⓓ
11) Ⓐ Ⓑ Ⓒ Ⓓ	51) Ⓐ Ⓑ Ⓒ Ⓓ	91) Ⓐ Ⓑ Ⓒ Ⓓ	131) Ⓐ Ⓑ Ⓒ Ⓓ
12) Ⓐ Ⓑ Ⓒ Ⓓ	52) Ⓐ Ⓑ Ⓒ Ⓓ	92) Ⓐ Ⓑ Ⓒ Ⓓ	132) Ⓐ Ⓑ Ⓒ Ⓓ
13) Ⓐ Ⓑ Ⓒ Ⓓ	53) Ⓐ Ⓑ Ⓒ Ⓓ	93) Ⓐ Ⓑ Ⓒ Ⓓ	133) Ⓐ Ⓑ Ⓒ Ⓓ
14) Ⓐ Ⓑ Ⓒ Ⓓ	54) Ⓐ Ⓑ Ⓒ Ⓓ	94) Ⓐ Ⓑ Ⓒ Ⓓ	134) Ⓐ Ⓑ Ⓒ Ⓓ
15) Ⓐ Ⓑ Ⓒ Ⓓ	55) Ⓐ Ⓑ Ⓒ Ⓓ	95) Ⓐ Ⓑ Ⓒ Ⓓ	135) Ⓐ Ⓑ Ⓒ Ⓓ
16) Ⓐ Ⓑ Ⓒ Ⓓ	56) Ⓐ Ⓑ Ⓒ Ⓓ	96) Ⓐ Ⓑ Ⓒ Ⓓ	136) Ⓐ Ⓑ Ⓒ Ⓓ
17) Ⓐ Ⓑ Ⓒ Ⓓ	57) Ⓐ Ⓑ Ⓒ Ⓓ	97) Ⓐ Ⓑ Ⓒ Ⓓ	137) Ⓐ Ⓑ Ⓒ Ⓓ
18) Ⓐ Ⓑ Ⓒ Ⓓ	58) Ⓐ Ⓑ Ⓒ Ⓓ	98) Ⓐ Ⓑ Ⓒ Ⓓ	138) Ⓐ Ⓑ Ⓒ Ⓓ
19) Ⓐ Ⓑ Ⓒ Ⓓ	59) Ⓐ Ⓑ Ⓒ Ⓓ	99) Ⓐ Ⓑ Ⓒ Ⓓ	139) Ⓐ Ⓑ Ⓒ Ⓓ
20) Ⓐ Ⓑ Ⓒ Ⓓ	60) Ⓐ Ⓑ Ⓒ Ⓓ	100) Ⓐ Ⓑ Ⓒ Ⓓ	140) Ⓐ Ⓑ Ⓒ Ⓓ
21) Ⓐ Ⓑ Ⓒ Ⓓ	61) Ⓐ Ⓑ Ⓒ Ⓓ	101) Ⓐ Ⓑ Ⓒ Ⓓ	141) Ⓐ Ⓑ Ⓒ Ⓓ
22) Ⓐ Ⓑ Ⓒ Ⓓ	62) Ⓐ Ⓑ Ⓒ Ⓓ	102) Ⓐ Ⓑ Ⓒ Ⓓ	142) Ⓐ Ⓑ Ⓒ Ⓓ
23) Ⓐ Ⓑ Ⓒ Ⓓ	63) Ⓐ Ⓑ Ⓒ Ⓓ	103) Ⓐ Ⓑ Ⓒ Ⓓ	143) Ⓐ Ⓑ Ⓒ Ⓓ
24) Ⓐ Ⓑ Ⓒ Ⓓ	64) Ⓐ Ⓑ Ⓒ Ⓓ	104) Ⓐ Ⓑ Ⓒ Ⓓ	144) Ⓐ Ⓑ Ⓒ Ⓓ
25) Ⓐ Ⓑ Ⓒ Ⓓ	65) Ⓐ Ⓑ Ⓒ Ⓓ	105) Ⓐ Ⓑ Ⓒ Ⓓ	145) Ⓐ Ⓑ Ⓒ Ⓓ
26) Ⓐ Ⓑ Ⓒ Ⓓ	66) Ⓐ Ⓑ Ⓒ Ⓓ	106) Ⓐ Ⓑ Ⓒ Ⓓ	146) Ⓐ Ⓑ Ⓒ Ⓓ
27) Ⓐ Ⓑ Ⓒ Ⓓ	67) Ⓐ Ⓑ Ⓒ Ⓓ	107) Ⓐ Ⓑ Ⓒ Ⓓ	147) Ⓐ Ⓑ Ⓒ Ⓓ
28) Ⓐ Ⓑ Ⓒ Ⓓ	68) Ⓐ Ⓑ Ⓒ Ⓓ	108) Ⓐ Ⓑ Ⓒ Ⓓ	148) Ⓐ Ⓑ Ⓒ Ⓓ
29) Ⓐ Ⓑ Ⓒ Ⓓ	69) Ⓐ Ⓑ Ⓒ Ⓓ	109) Ⓐ Ⓑ Ⓒ Ⓓ	149) Ⓐ Ⓑ Ⓒ Ⓓ
30) Ⓐ Ⓑ Ⓒ Ⓓ	70) Ⓐ Ⓑ Ⓒ Ⓓ	110) Ⓐ Ⓑ Ⓒ Ⓓ	150) Ⓐ Ⓑ Ⓒ Ⓓ
31) Ⓐ Ⓑ Ⓒ Ⓓ	71) Ⓐ Ⓑ Ⓒ Ⓓ	111) Ⓐ Ⓑ Ⓒ Ⓓ	
32) Ⓐ Ⓑ Ⓒ Ⓓ	72) Ⓐ Ⓑ Ⓒ Ⓓ	112) Ⓐ Ⓑ Ⓒ Ⓓ	
33) Ⓐ Ⓑ Ⓒ Ⓓ	73) Ⓐ Ⓑ Ⓒ Ⓓ	113) Ⓐ Ⓑ Ⓒ Ⓓ	
34) Ⓐ Ⓑ Ⓒ Ⓓ	74) Ⓐ Ⓑ Ⓒ Ⓓ	114) Ⓐ Ⓑ Ⓒ Ⓓ	
35) Ⓐ Ⓑ Ⓒ Ⓓ	75) Ⓐ Ⓑ Ⓒ Ⓓ	115) Ⓐ Ⓑ Ⓒ Ⓓ	
36) Ⓐ Ⓑ Ⓒ Ⓓ	76) Ⓐ Ⓑ Ⓒ Ⓓ	116) Ⓐ Ⓑ Ⓒ Ⓓ	
37) Ⓐ Ⓑ Ⓒ Ⓓ	77) Ⓐ Ⓑ Ⓒ Ⓓ	117) Ⓐ Ⓑ Ⓒ Ⓓ	
38) Ⓐ Ⓑ Ⓒ Ⓓ	78) Ⓐ Ⓑ Ⓒ Ⓓ	118) Ⓐ Ⓑ Ⓒ Ⓓ	
39) Ⓐ Ⓑ Ⓒ Ⓓ	79) Ⓐ Ⓑ Ⓒ Ⓓ	119) Ⓐ Ⓑ Ⓒ Ⓓ	
40) Ⓐ Ⓑ Ⓒ Ⓓ	80) Ⓐ Ⓑ Ⓒ Ⓓ	120) Ⓐ Ⓑ Ⓒ Ⓓ	

1) Ⓐ Ⓑ Ⓒ Ⓓ	41) Ⓐ Ⓑ Ⓒ Ⓓ	81) Ⓐ Ⓑ Ⓒ Ⓓ	121) Ⓐ Ⓑ Ⓒ Ⓓ
2) Ⓐ Ⓑ Ⓒ Ⓓ	42) Ⓐ Ⓑ Ⓒ Ⓓ	82) Ⓐ Ⓑ Ⓒ Ⓓ	122) Ⓐ Ⓑ Ⓒ Ⓓ
3) Ⓐ Ⓑ Ⓒ Ⓓ	43) Ⓐ Ⓑ Ⓒ Ⓓ	83) Ⓐ Ⓑ Ⓒ Ⓓ	123) Ⓐ Ⓑ Ⓒ Ⓓ
4) Ⓐ Ⓑ Ⓒ Ⓓ	44) Ⓐ Ⓑ Ⓒ Ⓓ	84) Ⓐ Ⓑ Ⓒ Ⓓ	124) Ⓐ Ⓑ Ⓒ Ⓓ
5) Ⓐ Ⓑ Ⓒ Ⓓ	45) Ⓐ Ⓑ Ⓒ Ⓓ	85) Ⓐ Ⓑ Ⓒ Ⓓ	125) Ⓐ Ⓑ Ⓒ Ⓓ
6) Ⓐ Ⓑ Ⓒ Ⓓ	46) Ⓐ Ⓑ Ⓒ Ⓓ	86) Ⓐ Ⓑ Ⓒ Ⓓ	126) Ⓐ Ⓑ Ⓒ Ⓓ
7) Ⓐ Ⓑ Ⓒ Ⓓ	47) Ⓐ Ⓑ Ⓒ Ⓓ	87) Ⓐ Ⓑ Ⓒ Ⓓ	127) Ⓐ Ⓑ Ⓒ Ⓓ
8) Ⓐ Ⓑ Ⓒ Ⓓ	48) Ⓐ Ⓑ Ⓒ Ⓓ	88) Ⓐ Ⓑ Ⓒ Ⓓ	128) Ⓐ Ⓑ Ⓒ Ⓓ
9) Ⓐ Ⓑ Ⓒ Ⓓ	49) Ⓐ Ⓑ Ⓒ Ⓓ	89) Ⓐ Ⓑ Ⓒ Ⓓ	129) Ⓐ Ⓑ Ⓒ Ⓓ
10) Ⓐ Ⓑ Ⓒ Ⓓ	50) Ⓐ Ⓑ Ⓒ Ⓓ	90) Ⓐ Ⓑ Ⓒ Ⓓ	130) Ⓐ Ⓑ Ⓒ Ⓓ
11) Ⓐ Ⓑ Ⓒ Ⓓ	51) Ⓐ Ⓑ Ⓒ Ⓓ	91) Ⓐ Ⓑ Ⓒ Ⓓ	131) Ⓐ Ⓑ Ⓒ Ⓓ
12) Ⓐ Ⓑ Ⓒ Ⓓ	52) Ⓐ Ⓑ Ⓒ Ⓓ	92) Ⓐ Ⓑ Ⓒ Ⓓ	132) Ⓐ Ⓑ Ⓒ Ⓓ
13) Ⓐ Ⓑ Ⓒ Ⓓ	53) Ⓐ Ⓑ Ⓒ Ⓓ	93) Ⓐ Ⓑ Ⓒ Ⓓ	133) Ⓐ Ⓑ Ⓒ Ⓓ
14) Ⓐ Ⓑ Ⓒ Ⓓ	54) Ⓐ Ⓑ Ⓒ Ⓓ	94) Ⓐ Ⓑ Ⓒ Ⓓ	134) Ⓐ Ⓑ Ⓒ Ⓓ
15) Ⓐ Ⓑ Ⓒ Ⓓ	55) Ⓐ Ⓑ Ⓒ Ⓓ	95) Ⓐ Ⓑ Ⓒ Ⓓ	135) Ⓐ Ⓑ Ⓒ Ⓓ
16) Ⓐ Ⓑ Ⓒ Ⓓ	56) Ⓐ Ⓑ Ⓒ Ⓓ	96) Ⓐ Ⓑ Ⓒ Ⓓ	136) Ⓐ Ⓑ Ⓒ Ⓓ
17) Ⓐ Ⓑ Ⓒ Ⓓ	57) Ⓐ Ⓑ Ⓒ Ⓓ	97) Ⓐ Ⓑ Ⓒ Ⓓ	137) Ⓐ Ⓑ Ⓒ Ⓓ
18) Ⓐ Ⓑ Ⓒ Ⓓ	58) Ⓐ Ⓑ Ⓒ Ⓓ	98) Ⓐ Ⓑ Ⓒ Ⓓ	138) Ⓐ Ⓑ Ⓒ Ⓓ
19) Ⓐ Ⓑ Ⓒ Ⓓ	59) Ⓐ Ⓑ Ⓒ Ⓓ	99) Ⓐ Ⓑ Ⓒ Ⓓ	139) Ⓐ Ⓑ Ⓒ Ⓓ
20) Ⓐ Ⓑ Ⓒ Ⓓ	60) Ⓐ Ⓑ Ⓒ Ⓓ	100) Ⓐ Ⓑ Ⓒ Ⓓ	140) Ⓐ Ⓑ Ⓒ Ⓓ
21) Ⓐ Ⓑ Ⓒ Ⓓ	61) Ⓐ Ⓑ Ⓒ Ⓓ	101) Ⓐ Ⓑ Ⓒ Ⓓ	141) Ⓐ Ⓑ Ⓒ Ⓓ
22) Ⓐ Ⓑ Ⓒ Ⓓ	62) Ⓐ Ⓑ Ⓒ Ⓓ	102) Ⓐ Ⓑ Ⓒ Ⓓ	142) Ⓐ Ⓑ Ⓒ Ⓓ
23) Ⓐ Ⓑ Ⓒ Ⓓ	63) Ⓐ Ⓑ Ⓒ Ⓓ	103) Ⓐ Ⓑ Ⓒ Ⓓ	143) Ⓐ Ⓑ Ⓒ Ⓓ
24) Ⓐ Ⓑ Ⓒ Ⓓ	64) Ⓐ Ⓑ Ⓒ Ⓓ	104) Ⓐ Ⓑ Ⓒ Ⓓ	144) Ⓐ Ⓑ Ⓒ Ⓓ
25) Ⓐ Ⓑ Ⓒ Ⓓ	65) Ⓐ Ⓑ Ⓒ Ⓓ	105) Ⓐ Ⓑ Ⓒ Ⓓ	145) Ⓐ Ⓑ Ⓒ Ⓓ
26) Ⓐ Ⓑ Ⓒ Ⓓ	66) Ⓐ Ⓑ Ⓒ Ⓓ	106) Ⓐ Ⓑ Ⓒ Ⓓ	146) Ⓐ Ⓑ Ⓒ Ⓓ
27) Ⓐ Ⓑ Ⓒ Ⓓ	67) Ⓐ Ⓑ Ⓒ Ⓓ	107) Ⓐ Ⓑ Ⓒ Ⓓ	147) Ⓐ Ⓑ Ⓒ Ⓓ
28) Ⓐ Ⓑ Ⓒ Ⓓ	68) Ⓐ Ⓑ Ⓒ Ⓓ	108) Ⓐ Ⓑ Ⓒ Ⓓ	148) Ⓐ Ⓑ Ⓒ Ⓓ
29) Ⓐ Ⓑ Ⓒ Ⓓ	69) Ⓐ Ⓑ Ⓒ Ⓓ	109) Ⓐ Ⓑ Ⓒ Ⓓ	149) Ⓐ Ⓑ Ⓒ Ⓓ
30) Ⓐ Ⓑ Ⓒ Ⓓ	70) Ⓐ Ⓑ Ⓒ Ⓓ	110) Ⓐ Ⓑ Ⓒ Ⓓ	150) Ⓐ Ⓑ Ⓒ Ⓓ
31) Ⓐ Ⓑ Ⓒ Ⓓ	71) Ⓐ Ⓑ Ⓒ Ⓓ	111) Ⓐ Ⓑ Ⓒ Ⓓ	
32) Ⓐ Ⓑ Ⓒ Ⓓ	72) Ⓐ Ⓑ Ⓒ Ⓓ	112) Ⓐ Ⓑ Ⓒ Ⓓ	
33) Ⓐ Ⓑ Ⓒ Ⓓ	73) Ⓐ Ⓑ Ⓒ Ⓓ	113) Ⓐ Ⓑ Ⓒ Ⓓ	
34) Ⓐ Ⓑ Ⓒ Ⓓ	74) Ⓐ Ⓑ Ⓒ Ⓓ	114) Ⓐ Ⓑ Ⓒ Ⓓ	
35) Ⓐ Ⓑ Ⓒ Ⓓ	75) Ⓐ Ⓑ Ⓒ Ⓓ	115) Ⓐ Ⓑ Ⓒ Ⓓ	
36) Ⓐ Ⓑ Ⓒ Ⓓ	76) Ⓐ Ⓑ Ⓒ Ⓓ	116) Ⓐ Ⓑ Ⓒ Ⓓ	
37) Ⓐ Ⓑ Ⓒ Ⓓ	77) Ⓐ Ⓑ Ⓒ Ⓓ	117) Ⓐ Ⓑ Ⓒ Ⓓ	
38) Ⓐ Ⓑ Ⓒ Ⓓ	78) Ⓐ Ⓑ Ⓒ Ⓓ	118) Ⓐ Ⓑ Ⓒ Ⓓ	
39) Ⓐ Ⓑ Ⓒ Ⓓ	79) Ⓐ Ⓑ Ⓒ Ⓓ	119) Ⓐ Ⓑ Ⓒ Ⓓ	
40) Ⓐ Ⓑ Ⓒ Ⓓ	80) Ⓐ Ⓑ Ⓒ Ⓓ	120) Ⓐ Ⓑ Ⓒ Ⓓ	

1) Ⓐ Ⓑ Ⓒ Ⓓ	41) Ⓐ Ⓑ Ⓒ Ⓓ	81) Ⓐ Ⓑ Ⓒ Ⓓ	121) Ⓐ Ⓑ Ⓒ Ⓓ
2) Ⓐ Ⓑ Ⓒ Ⓓ	42) Ⓐ Ⓑ Ⓒ Ⓓ	82) Ⓐ Ⓑ Ⓒ Ⓓ	122) Ⓐ Ⓑ Ⓒ Ⓓ
3) Ⓐ Ⓑ Ⓒ Ⓓ	43) Ⓐ Ⓑ Ⓒ Ⓓ	83) Ⓐ Ⓑ Ⓒ Ⓓ	123) Ⓐ Ⓑ Ⓒ Ⓓ
4) Ⓐ Ⓑ Ⓒ Ⓓ	44) Ⓐ Ⓑ Ⓒ Ⓓ	84) Ⓐ Ⓑ Ⓒ Ⓓ	124) Ⓐ Ⓑ Ⓒ Ⓓ
5) Ⓐ Ⓑ Ⓒ Ⓓ	45) Ⓐ Ⓑ Ⓒ Ⓓ	85) Ⓐ Ⓑ Ⓒ Ⓓ	125) Ⓐ Ⓑ Ⓒ Ⓓ
6) Ⓐ Ⓑ Ⓒ Ⓓ	46) Ⓐ Ⓑ Ⓒ Ⓓ	86) Ⓐ Ⓑ Ⓒ Ⓓ	126) Ⓐ Ⓑ Ⓒ Ⓓ
7) Ⓐ Ⓑ Ⓒ Ⓓ	47) Ⓐ Ⓑ Ⓒ Ⓓ	87) Ⓐ Ⓑ Ⓒ Ⓓ	127) Ⓐ Ⓑ Ⓒ Ⓓ
8) Ⓐ Ⓑ Ⓒ Ⓓ	48) Ⓐ Ⓑ Ⓒ Ⓓ	88) Ⓐ Ⓑ Ⓒ Ⓓ	128) Ⓐ Ⓑ Ⓒ Ⓓ
9) Ⓐ Ⓑ Ⓒ Ⓓ	49) Ⓐ Ⓑ Ⓒ Ⓓ	89) Ⓐ Ⓑ Ⓒ Ⓓ	129) Ⓐ Ⓑ Ⓒ Ⓓ
10) Ⓐ Ⓑ Ⓒ Ⓓ	50) Ⓐ Ⓑ Ⓒ Ⓓ	90) Ⓐ Ⓑ Ⓒ Ⓓ	130) Ⓐ Ⓑ Ⓒ Ⓓ
11) Ⓐ Ⓑ Ⓒ Ⓓ	51) Ⓐ Ⓑ Ⓒ Ⓓ	91) Ⓐ Ⓑ Ⓒ Ⓓ	131) Ⓐ Ⓑ Ⓒ Ⓓ
12) Ⓐ Ⓑ Ⓒ Ⓓ	52) Ⓐ Ⓑ Ⓒ Ⓓ	92) Ⓐ Ⓑ Ⓒ Ⓓ	132) Ⓐ Ⓑ Ⓒ Ⓓ
13) Ⓐ Ⓑ Ⓒ Ⓓ	53) Ⓐ Ⓑ Ⓒ Ⓓ	93) Ⓐ Ⓑ Ⓒ Ⓓ	133) Ⓐ Ⓑ Ⓒ Ⓓ
14) Ⓐ Ⓑ Ⓒ Ⓓ	54) Ⓐ Ⓑ Ⓒ Ⓓ	94) Ⓐ Ⓑ Ⓒ Ⓓ	134) Ⓐ Ⓑ Ⓒ Ⓓ
15) Ⓐ Ⓑ Ⓒ Ⓓ	55) Ⓐ Ⓑ Ⓒ Ⓓ	95) Ⓐ Ⓑ Ⓒ Ⓓ	135) Ⓐ Ⓑ Ⓒ Ⓓ
16) Ⓐ Ⓑ Ⓒ Ⓓ	56) Ⓐ Ⓑ Ⓒ Ⓓ	96) Ⓐ Ⓑ Ⓒ Ⓓ	136) Ⓐ Ⓑ Ⓒ Ⓓ
17) Ⓐ Ⓑ Ⓒ Ⓓ	57) Ⓐ Ⓑ Ⓒ Ⓓ	97) Ⓐ Ⓑ Ⓒ Ⓓ	137) Ⓐ Ⓑ Ⓒ Ⓓ
18) Ⓐ Ⓑ Ⓒ Ⓓ	58) Ⓐ Ⓑ Ⓒ Ⓓ	98) Ⓐ Ⓑ Ⓒ Ⓓ	138) Ⓐ Ⓑ Ⓒ Ⓓ
19) Ⓐ Ⓑ Ⓒ Ⓓ	59) Ⓐ Ⓑ Ⓒ Ⓓ	99) Ⓐ Ⓑ Ⓒ Ⓓ	139) Ⓐ Ⓑ Ⓒ Ⓓ
20) Ⓐ Ⓑ Ⓒ Ⓓ	60) Ⓐ Ⓑ Ⓒ Ⓓ	100) Ⓐ Ⓑ Ⓒ Ⓓ	140) Ⓐ Ⓑ Ⓒ Ⓓ
21) Ⓐ Ⓑ Ⓒ Ⓓ	61) Ⓐ Ⓑ Ⓒ Ⓓ	101) Ⓐ Ⓑ Ⓒ Ⓓ	141) Ⓐ Ⓑ Ⓒ Ⓓ
22) Ⓐ Ⓑ Ⓒ Ⓓ	62) Ⓐ Ⓑ Ⓒ Ⓓ	102) Ⓐ Ⓑ Ⓒ Ⓓ	142) Ⓐ Ⓑ Ⓒ Ⓓ
23) Ⓐ Ⓑ Ⓒ Ⓓ	63) Ⓐ Ⓑ Ⓒ Ⓓ	103) Ⓐ Ⓑ Ⓒ Ⓓ	143) Ⓐ Ⓑ Ⓒ Ⓓ
24) Ⓐ Ⓑ Ⓒ Ⓓ	64) Ⓐ Ⓑ Ⓒ Ⓓ	104) Ⓐ Ⓑ Ⓒ Ⓓ	144) Ⓐ Ⓑ Ⓒ Ⓓ
25) Ⓐ Ⓑ Ⓒ Ⓓ	65) Ⓐ Ⓑ Ⓒ Ⓓ	105) Ⓐ Ⓑ Ⓒ Ⓓ	145) Ⓐ Ⓑ Ⓒ Ⓓ
26) Ⓐ Ⓑ Ⓒ Ⓓ	66) Ⓐ Ⓑ Ⓒ Ⓓ	106) Ⓐ Ⓑ Ⓒ Ⓓ	146) Ⓐ Ⓑ Ⓒ Ⓓ
27) Ⓐ Ⓑ Ⓒ Ⓓ	67) Ⓐ Ⓑ Ⓒ Ⓓ	107) Ⓐ Ⓑ Ⓒ Ⓓ	147) Ⓐ Ⓑ Ⓒ Ⓓ
28) Ⓐ Ⓑ Ⓒ Ⓓ	68) Ⓐ Ⓑ Ⓒ Ⓓ	108) Ⓐ Ⓑ Ⓒ Ⓓ	148) Ⓐ Ⓑ Ⓒ Ⓓ
29) Ⓐ Ⓑ Ⓒ Ⓓ	69) Ⓐ Ⓑ Ⓒ Ⓓ	109) Ⓐ Ⓑ Ⓒ Ⓓ	149) Ⓐ Ⓑ Ⓒ Ⓓ
30) Ⓐ Ⓑ Ⓒ Ⓓ	70) Ⓐ Ⓑ Ⓒ Ⓓ	110) Ⓐ Ⓑ Ⓒ Ⓓ	150) Ⓐ Ⓑ Ⓒ Ⓓ
31) Ⓐ Ⓑ Ⓒ Ⓓ	71) Ⓐ Ⓑ Ⓒ Ⓓ	111) Ⓐ Ⓑ Ⓒ Ⓓ	
32) Ⓐ Ⓑ Ⓒ Ⓓ	72) Ⓐ Ⓑ Ⓒ Ⓓ	112) Ⓐ Ⓑ Ⓒ Ⓓ	
33) Ⓐ Ⓑ Ⓒ Ⓓ	73) Ⓐ Ⓑ Ⓒ Ⓓ	113) Ⓐ Ⓑ Ⓒ Ⓓ	
34) Ⓐ Ⓑ Ⓒ Ⓓ	74) Ⓐ Ⓑ Ⓒ Ⓓ	114) Ⓐ Ⓑ Ⓒ Ⓓ	
35) Ⓐ Ⓑ Ⓒ Ⓓ	75) Ⓐ Ⓑ Ⓒ Ⓓ	115) Ⓐ Ⓑ Ⓒ Ⓓ	
36) Ⓐ Ⓑ Ⓒ Ⓓ	76) Ⓐ Ⓑ Ⓒ Ⓓ	116) Ⓐ Ⓑ Ⓒ Ⓓ	
37) Ⓐ Ⓑ Ⓒ Ⓓ	77) Ⓐ Ⓑ Ⓒ Ⓓ	117) Ⓐ Ⓑ Ⓒ Ⓓ	
38) Ⓐ Ⓑ Ⓒ Ⓓ	78) Ⓐ Ⓑ Ⓒ Ⓓ	118) Ⓐ Ⓑ Ⓒ Ⓓ	
39) Ⓐ Ⓑ Ⓒ Ⓓ	79) Ⓐ Ⓑ Ⓒ Ⓓ	119) Ⓐ Ⓑ Ⓒ Ⓓ	
40) Ⓐ Ⓑ Ⓒ Ⓓ	80) Ⓐ Ⓑ Ⓒ Ⓓ	120) Ⓐ Ⓑ Ⓒ Ⓓ	

1) Ⓐ Ⓑ Ⓒ Ⓓ	41) Ⓐ Ⓑ Ⓒ Ⓓ	81) Ⓐ Ⓑ Ⓒ Ⓓ	121) Ⓐ Ⓑ Ⓒ Ⓓ
2) Ⓐ Ⓑ Ⓒ Ⓓ	42) Ⓐ Ⓑ Ⓒ Ⓓ	82) Ⓐ Ⓑ Ⓒ Ⓓ	122) Ⓐ Ⓑ Ⓒ Ⓓ
3) Ⓐ Ⓑ Ⓒ Ⓓ	43) Ⓐ Ⓑ Ⓒ Ⓓ	83) Ⓐ Ⓑ Ⓒ Ⓓ	123) Ⓐ Ⓑ Ⓒ Ⓓ
4) Ⓐ Ⓑ Ⓒ Ⓓ	44) Ⓐ Ⓑ Ⓒ Ⓓ	84) Ⓐ Ⓑ Ⓒ Ⓓ	124) Ⓐ Ⓑ Ⓒ Ⓓ
5) Ⓐ Ⓑ Ⓒ Ⓓ	45) Ⓐ Ⓑ Ⓒ Ⓓ	85) Ⓐ Ⓑ Ⓒ Ⓓ	125) Ⓐ Ⓑ Ⓒ Ⓓ
6) Ⓐ Ⓑ Ⓒ Ⓓ	46) Ⓐ Ⓑ Ⓒ Ⓓ	86) Ⓐ Ⓑ Ⓒ Ⓓ	126) Ⓐ Ⓑ Ⓒ Ⓓ
7) Ⓐ Ⓑ Ⓒ Ⓓ	47) Ⓐ Ⓑ Ⓒ Ⓓ	87) Ⓐ Ⓑ Ⓒ Ⓓ	127) Ⓐ Ⓑ Ⓒ Ⓓ
8) Ⓐ Ⓑ Ⓒ Ⓓ	48) Ⓐ Ⓑ Ⓒ Ⓓ	88) Ⓐ Ⓑ Ⓒ Ⓓ	128) Ⓐ Ⓑ Ⓒ Ⓓ
9) Ⓐ Ⓑ Ⓒ Ⓓ	49) Ⓐ Ⓑ Ⓒ Ⓓ	89) Ⓐ Ⓑ Ⓒ Ⓓ	129) Ⓐ Ⓑ Ⓒ Ⓓ
10) Ⓐ Ⓑ Ⓒ Ⓓ	50) Ⓐ Ⓑ Ⓒ Ⓓ	90) Ⓐ Ⓑ Ⓒ Ⓓ	130) Ⓐ Ⓑ Ⓒ Ⓓ
11) Ⓐ Ⓑ Ⓒ Ⓓ	51) Ⓐ Ⓑ Ⓒ Ⓓ	91) Ⓐ Ⓑ Ⓒ Ⓓ	131) Ⓐ Ⓑ Ⓒ Ⓓ
12) Ⓐ Ⓑ Ⓒ Ⓓ	52) Ⓐ Ⓑ Ⓒ Ⓓ	92) Ⓐ Ⓑ Ⓒ Ⓓ	132) Ⓐ Ⓑ Ⓒ Ⓓ
13) Ⓐ Ⓑ Ⓒ Ⓓ	53) Ⓐ Ⓑ Ⓒ Ⓓ	93) Ⓐ Ⓑ Ⓒ Ⓓ	133) Ⓐ Ⓑ Ⓒ Ⓓ
14) Ⓐ Ⓑ Ⓒ Ⓓ	54) Ⓐ Ⓑ Ⓒ Ⓓ	94) Ⓐ Ⓑ Ⓒ Ⓓ	134) Ⓐ Ⓑ Ⓒ Ⓓ
15) Ⓐ Ⓑ Ⓒ Ⓓ	55) Ⓐ Ⓑ Ⓒ Ⓓ	95) Ⓐ Ⓑ Ⓒ Ⓓ	135) Ⓐ Ⓑ Ⓒ Ⓓ
16) Ⓐ Ⓑ Ⓒ Ⓓ	56) Ⓐ Ⓑ Ⓒ Ⓓ	96) Ⓐ Ⓑ Ⓒ Ⓓ	136) Ⓐ Ⓑ Ⓒ Ⓓ
17) Ⓐ Ⓑ Ⓒ Ⓓ	57) Ⓐ Ⓑ Ⓒ Ⓓ	97) Ⓐ Ⓑ Ⓒ Ⓓ	137) Ⓐ Ⓑ Ⓒ Ⓓ
18) Ⓐ Ⓑ Ⓒ Ⓓ	58) Ⓐ Ⓑ Ⓒ Ⓓ	98) Ⓐ Ⓑ Ⓒ Ⓓ	138) Ⓐ Ⓑ Ⓒ Ⓓ
19) Ⓐ Ⓑ Ⓒ Ⓓ	59) Ⓐ Ⓑ Ⓒ Ⓓ	99) Ⓐ Ⓑ Ⓒ Ⓓ	139) Ⓐ Ⓑ Ⓒ Ⓓ
20) Ⓐ Ⓑ Ⓒ Ⓓ	60) Ⓐ Ⓑ Ⓒ Ⓓ	100) Ⓐ Ⓑ Ⓒ Ⓓ	140) Ⓐ Ⓑ Ⓒ Ⓓ
21) Ⓐ Ⓑ Ⓒ Ⓓ	61) Ⓐ Ⓑ Ⓒ Ⓓ	101) Ⓐ Ⓑ Ⓒ Ⓓ	141) Ⓐ Ⓑ Ⓒ Ⓓ
22) Ⓐ Ⓑ Ⓒ Ⓓ	62) Ⓐ Ⓑ Ⓒ Ⓓ	102) Ⓐ Ⓑ Ⓒ Ⓓ	142) Ⓐ Ⓑ Ⓒ Ⓓ
23) Ⓐ Ⓑ Ⓒ Ⓓ	63) Ⓐ Ⓑ Ⓒ Ⓓ	103) Ⓐ Ⓑ Ⓒ Ⓓ	143) Ⓐ Ⓑ Ⓒ Ⓓ
24) Ⓐ Ⓑ Ⓒ Ⓓ	64) Ⓐ Ⓑ Ⓒ Ⓓ	104) Ⓐ Ⓑ Ⓒ Ⓓ	144) Ⓐ Ⓑ Ⓒ Ⓓ
25) Ⓐ Ⓑ Ⓒ Ⓓ	65) Ⓐ Ⓑ Ⓒ Ⓓ	105) Ⓐ Ⓑ Ⓒ Ⓓ	145) Ⓐ Ⓑ Ⓒ Ⓓ
26) Ⓐ Ⓑ Ⓒ Ⓓ	66) Ⓐ Ⓑ Ⓒ Ⓓ	106) Ⓐ Ⓑ Ⓒ Ⓓ	146) Ⓐ Ⓑ Ⓒ Ⓓ
27) Ⓐ Ⓑ Ⓒ Ⓓ	67) Ⓐ Ⓑ Ⓒ Ⓓ	107) Ⓐ Ⓑ Ⓒ Ⓓ	147) Ⓐ Ⓑ Ⓒ Ⓓ
28) Ⓐ Ⓑ Ⓒ Ⓓ	68) Ⓐ Ⓑ Ⓒ Ⓓ	108) Ⓐ Ⓑ Ⓒ Ⓓ	148) Ⓐ Ⓑ Ⓒ Ⓓ
29) Ⓐ Ⓑ Ⓒ Ⓓ	69) Ⓐ Ⓑ Ⓒ Ⓓ	109) Ⓐ Ⓑ Ⓒ Ⓓ	149) Ⓐ Ⓑ Ⓒ Ⓓ
30) Ⓐ Ⓑ Ⓒ Ⓓ	70) Ⓐ Ⓑ Ⓒ Ⓓ	110) Ⓐ Ⓑ Ⓒ Ⓓ	150) Ⓐ Ⓑ Ⓒ Ⓓ
31) Ⓐ Ⓑ Ⓒ Ⓓ	71) Ⓐ Ⓑ Ⓒ Ⓓ	111) Ⓐ Ⓑ Ⓒ Ⓓ	
32) Ⓐ Ⓑ Ⓒ Ⓓ	72) Ⓐ Ⓑ Ⓒ Ⓓ	112) Ⓐ Ⓑ Ⓒ Ⓓ	
33) Ⓐ Ⓑ Ⓒ Ⓓ	73) Ⓐ Ⓑ Ⓒ Ⓓ	113) Ⓐ Ⓑ Ⓒ Ⓓ	
34) Ⓐ Ⓑ Ⓒ Ⓓ	74) Ⓐ Ⓑ Ⓒ Ⓓ	114) Ⓐ Ⓑ Ⓒ Ⓓ	
35) Ⓐ Ⓑ Ⓒ Ⓓ	75) Ⓐ Ⓑ Ⓒ Ⓓ	115) Ⓐ Ⓑ Ⓒ Ⓓ	
36) Ⓐ Ⓑ Ⓒ Ⓓ	76) Ⓐ Ⓑ Ⓒ Ⓓ	116) Ⓐ Ⓑ Ⓒ Ⓓ	
37) Ⓐ Ⓑ Ⓒ Ⓓ	77) Ⓐ Ⓑ Ⓒ Ⓓ	117) Ⓐ Ⓑ Ⓒ Ⓓ	
38) Ⓐ Ⓑ Ⓒ Ⓓ	78) Ⓐ Ⓑ Ⓒ Ⓓ	118) Ⓐ Ⓑ Ⓒ Ⓓ	
39) Ⓐ Ⓑ Ⓒ Ⓓ	79) Ⓐ Ⓑ Ⓒ Ⓓ	119) Ⓐ Ⓑ Ⓒ Ⓓ	
40) Ⓐ Ⓑ Ⓒ Ⓓ	80) Ⓐ Ⓑ Ⓒ Ⓓ	120) Ⓐ Ⓑ Ⓒ Ⓓ	

1) Ⓐ Ⓑ Ⓒ Ⓓ	41) Ⓐ Ⓑ Ⓒ Ⓓ	81) Ⓐ Ⓑ Ⓒ Ⓓ	121) Ⓐ Ⓑ Ⓒ Ⓓ
2) Ⓐ Ⓑ Ⓒ Ⓓ	42) Ⓐ Ⓑ Ⓒ Ⓓ	82) Ⓐ Ⓑ Ⓒ Ⓓ	122) Ⓐ Ⓑ Ⓒ Ⓓ
3) Ⓐ Ⓑ Ⓒ Ⓓ	43) Ⓐ Ⓑ Ⓒ Ⓓ	83) Ⓐ Ⓑ Ⓒ Ⓓ	123) Ⓐ Ⓑ Ⓒ Ⓓ
4) Ⓐ Ⓑ Ⓒ Ⓓ	44) Ⓐ Ⓑ Ⓒ Ⓓ	84) Ⓐ Ⓑ Ⓒ Ⓓ	124) Ⓐ Ⓑ Ⓒ Ⓓ
5) Ⓐ Ⓑ Ⓒ Ⓓ	45) Ⓐ Ⓑ Ⓒ Ⓓ	85) Ⓐ Ⓑ Ⓒ Ⓓ	125) Ⓐ Ⓑ Ⓒ Ⓓ
6) Ⓐ Ⓑ Ⓒ Ⓓ	46) Ⓐ Ⓑ Ⓒ Ⓓ	86) Ⓐ Ⓑ Ⓒ Ⓓ	126) Ⓐ Ⓑ Ⓒ Ⓓ
7) Ⓐ Ⓑ Ⓒ Ⓓ	47) Ⓐ Ⓑ Ⓒ Ⓓ	87) Ⓐ Ⓑ Ⓒ Ⓓ	127) Ⓐ Ⓑ Ⓒ Ⓓ
8) Ⓐ Ⓑ Ⓒ Ⓓ	48) Ⓐ Ⓑ Ⓒ Ⓓ	88) Ⓐ Ⓑ Ⓒ Ⓓ	128) Ⓐ Ⓑ Ⓒ Ⓓ
9) Ⓐ Ⓑ Ⓒ Ⓓ	49) Ⓐ Ⓑ Ⓒ Ⓓ	89) Ⓐ Ⓑ Ⓒ Ⓓ	129) Ⓐ Ⓑ Ⓒ Ⓓ
10) Ⓐ Ⓑ Ⓒ Ⓓ	50) Ⓐ Ⓑ Ⓒ Ⓓ	90) Ⓐ Ⓑ Ⓒ Ⓓ	130) Ⓐ Ⓑ Ⓒ Ⓓ
11) Ⓐ Ⓑ Ⓒ Ⓓ	51) Ⓐ Ⓑ Ⓒ Ⓓ	91) Ⓐ Ⓑ Ⓒ Ⓓ	131) Ⓐ Ⓑ Ⓒ Ⓓ
12) Ⓐ Ⓑ Ⓒ Ⓓ	52) Ⓐ Ⓑ Ⓒ Ⓓ	92) Ⓐ Ⓑ Ⓒ Ⓓ	132) Ⓐ Ⓑ Ⓒ Ⓓ
13) Ⓐ Ⓑ Ⓒ Ⓓ	53) Ⓐ Ⓑ Ⓒ Ⓓ	93) Ⓐ Ⓑ Ⓒ Ⓓ	133) Ⓐ Ⓑ Ⓒ Ⓓ
14) Ⓐ Ⓑ Ⓒ Ⓓ	54) Ⓐ Ⓑ Ⓒ Ⓓ	94) Ⓐ Ⓑ Ⓒ Ⓓ	134) Ⓐ Ⓑ Ⓒ Ⓓ
15) Ⓐ Ⓑ Ⓒ Ⓓ	55) Ⓐ Ⓑ Ⓒ Ⓓ	95) Ⓐ Ⓑ Ⓒ Ⓓ	135) Ⓐ Ⓑ Ⓒ Ⓓ
16) Ⓐ Ⓑ Ⓒ Ⓓ	56) Ⓐ Ⓑ Ⓒ Ⓓ	96) Ⓐ Ⓑ Ⓒ Ⓓ	136) Ⓐ Ⓑ Ⓒ Ⓓ
17) Ⓐ Ⓑ Ⓒ Ⓓ	57) Ⓐ Ⓑ Ⓒ Ⓓ	97) Ⓐ Ⓑ Ⓒ Ⓓ	137) Ⓐ Ⓑ Ⓒ Ⓓ
18) Ⓐ Ⓑ Ⓒ Ⓓ	58) Ⓐ Ⓑ Ⓒ Ⓓ	98) Ⓐ Ⓑ Ⓒ Ⓓ	138) Ⓐ Ⓑ Ⓒ Ⓓ
19) Ⓐ Ⓑ Ⓒ Ⓓ	59) Ⓐ Ⓑ Ⓒ Ⓓ	99) Ⓐ Ⓑ Ⓒ Ⓓ	139) Ⓐ Ⓑ Ⓒ Ⓓ
20) Ⓐ Ⓑ Ⓒ Ⓓ	60) Ⓐ Ⓑ Ⓒ Ⓓ	100) Ⓐ Ⓑ Ⓒ Ⓓ	140) Ⓐ Ⓑ Ⓒ Ⓓ
21) Ⓐ Ⓑ Ⓒ Ⓓ	61) Ⓐ Ⓑ Ⓒ Ⓓ	101) Ⓐ Ⓑ Ⓒ Ⓓ	141) Ⓐ Ⓑ Ⓒ Ⓓ
22) Ⓐ Ⓑ Ⓒ Ⓓ	62) Ⓐ Ⓑ Ⓒ Ⓓ	102) Ⓐ Ⓑ Ⓒ Ⓓ	142) Ⓐ Ⓑ Ⓒ Ⓓ
23) Ⓐ Ⓑ Ⓒ Ⓓ	63) Ⓐ Ⓑ Ⓒ Ⓓ	103) Ⓐ Ⓑ Ⓒ Ⓓ	143) Ⓐ Ⓑ Ⓒ Ⓓ
24) Ⓐ Ⓑ Ⓒ Ⓓ	64) Ⓐ Ⓑ Ⓒ Ⓓ	104) Ⓐ Ⓑ Ⓒ Ⓓ	144) Ⓐ Ⓑ Ⓒ Ⓓ
25) Ⓐ Ⓑ Ⓒ Ⓓ	65) Ⓐ Ⓑ Ⓒ Ⓓ	105) Ⓐ Ⓑ Ⓒ Ⓓ	145) Ⓐ Ⓑ Ⓒ Ⓓ
26) Ⓐ Ⓑ Ⓒ Ⓓ	66) Ⓐ Ⓑ Ⓒ Ⓓ	106) Ⓐ Ⓑ Ⓒ Ⓓ	146) Ⓐ Ⓑ Ⓒ Ⓓ
27) Ⓐ Ⓑ Ⓒ Ⓓ	67) Ⓐ Ⓑ Ⓒ Ⓓ	107) Ⓐ Ⓑ Ⓒ Ⓓ	147) Ⓐ Ⓑ Ⓒ Ⓓ
28) Ⓐ Ⓑ Ⓒ Ⓓ	68) Ⓐ Ⓑ Ⓒ Ⓓ	108) Ⓐ Ⓑ Ⓒ Ⓓ	148) Ⓐ Ⓑ Ⓒ Ⓓ
29) Ⓐ Ⓑ Ⓒ Ⓓ	69) Ⓐ Ⓑ Ⓒ Ⓓ	109) Ⓐ Ⓑ Ⓒ Ⓓ	149) Ⓐ Ⓑ Ⓒ Ⓓ
30) Ⓐ Ⓑ Ⓒ Ⓓ	70) Ⓐ Ⓑ Ⓒ Ⓓ	110) Ⓐ Ⓑ Ⓒ Ⓓ	150) Ⓐ Ⓑ Ⓒ Ⓓ
31) Ⓐ Ⓑ Ⓒ Ⓓ	71) Ⓐ Ⓑ Ⓒ Ⓓ	111) Ⓐ Ⓑ Ⓒ Ⓓ	
32) Ⓐ Ⓑ Ⓒ Ⓓ	72) Ⓐ Ⓑ Ⓒ Ⓓ	112) Ⓐ Ⓑ Ⓒ Ⓓ	
33) Ⓐ Ⓑ Ⓒ Ⓓ	73) Ⓐ Ⓑ Ⓒ Ⓓ	113) Ⓐ Ⓑ Ⓒ Ⓓ	
34) Ⓐ Ⓑ Ⓒ Ⓓ	74) Ⓐ Ⓑ Ⓒ Ⓓ	114) Ⓐ Ⓑ Ⓒ Ⓓ	
35) Ⓐ Ⓑ Ⓒ Ⓓ	75) Ⓐ Ⓑ Ⓒ Ⓓ	115) Ⓐ Ⓑ Ⓒ Ⓓ	
36) Ⓐ Ⓑ Ⓒ Ⓓ	76) Ⓐ Ⓑ Ⓒ Ⓓ	116) Ⓐ Ⓑ Ⓒ Ⓓ	
37) Ⓐ Ⓑ Ⓒ Ⓓ	77) Ⓐ Ⓑ Ⓒ Ⓓ	117) Ⓐ Ⓑ Ⓒ Ⓓ	
38) Ⓐ Ⓑ Ⓒ Ⓓ	78) Ⓐ Ⓑ Ⓒ Ⓓ	118) Ⓐ Ⓑ Ⓒ Ⓓ	
39) Ⓐ Ⓑ Ⓒ Ⓓ	79) Ⓐ Ⓑ Ⓒ Ⓓ	119) Ⓐ Ⓑ Ⓒ Ⓓ	
40) Ⓐ Ⓑ Ⓒ Ⓓ	80) Ⓐ Ⓑ Ⓒ Ⓓ	120) Ⓐ Ⓑ Ⓒ Ⓓ	